BIOHACKED

CHINA'S RACE TO CONTROL LIFE

Brandon J. Weichert

Foreword by Gordon G. Chang

BOOKS

New York · London

CONTENTS

FOREWORD
CHINA'S CIVILIZATION-KILLING MACHINE

GORDON G. CHANG[1]

COVID-19 "came out of the box ready to infect." So said Dr. Deborah Birx, President Trump's coronavirus task force coordinator, to London's *Mail on Sunday* in July 2022.

SARS-CoV-2, the pathogen causing this disease, "was already more infectious than flu when it first arrived," Birx noted. Most viruses take years to become highly transmissible.

"In laboratories you grow the virus in human cells, allowing it to adapt more," she told the paper. "Each time it passes through human cells it becomes more adapted."

Every crime leaves a clue. Perhaps the most damning clue of China's crime is the fast human transmission of SARS-CoV-2. Birx's comment follows those of Alina Chan, a microbiologist at the Broad Institute of Harvard and MIT, who angered the world's scientific community in 2020 by arguing that the startling lack of mutations in SARS-CoV-2 was evidence that it had been developed in a lab. If there had been zoonotic transfers of the virus from animals to humans – the widely accepted theory at the time – there should have been evolutions of the pathogen as it adapted to humans. Surprisingly, there was an unusual stability of the virus despite trillions of replications.

There are other clues. The unnatural features of the virus suggest gene splicing: the irregular arrangements of amino acids and the much-discussed furin cleavage site, for instance. Additionally, no one has found the natural reservoir of SARS-CoV-2 or been able to document the transmission links as it supposedly made its way from creatures to humans. Most tellingly, the Chinese regime from the beginning of the outbreak to today has gone to great lengths, including destroying samples and other evidence, to hinder the international community from investigating the origins of the disease.

There is only one conclusion fitting the facts: SARS-CoV-2 was manufactured in a Chinese facility.

Bill Gates, who has become a health campaigner of sorts, in 2022 said another great disease is on the way, and he talked about "bioterrorism" as a possible cause. Gates is right to be concerned because the Chinese state has already unleashed a biological weapon that has killed millions outside China.

Even if SARS-CoV-2 was the result of a natural process, Chinese leaders turned it into a weapon. Since at least December 2019, China's officials lied about the human-to-human transmissibility of the pathogen, telling the world it was not readily contagious when they knew it was highly so. Beijing propagated untruths through the statements of the World Health Organization and in China's direct communications with the health authorities of other nations.

The Chinese, as we learned from public comments of Birx and others in 2020, in fact lulled countries into not taking precautions. In her March 31 press briefing that year, for example, she said that after seeing Chinese data she thought the novel coronavirus was no more serious than SARS, the 2002 outbreak in China that eventually killed fewer than 800 people worldwide. It was only after witnessing the devastation in Italy and Spain that Birx realized she had been misled. Dr. Anthony Fauci, who headed the National Institute of Allergy and Infectious Diseases, has also publicly talked about being deceived by the Chinese.

Moreover, while Chinese leaders were locking down Wuhan and other cities – the lockdowns indicated they believed this tactic was effective in stopping transmission – they were pressuring other countries to accept arrivals from China without restriction or quarantine. The Communist Party had to know it was, with these actions, spreading the disease. This means the millions of deaths outside China were intentional – in other words, murder.

Is Chinese communism really that vicious?

The Communist Party uses the Soviet empirical framework called Comprehensive National Power to rank the strength of countries. China wants the No. 1 ranking, and there are two ways to obtain it: increase China's power or decrease everybody else's. And that's what Chinese officials did: they used disease to weaken others once COVID-19 had infected China.

Chinese leaders are prepared to devastate all other societies. The country's researchers even discuss bioweapons in public journals. China's National Defense University, in the 2017 edition of the authoritative *Science of Military Strategy*, mentioned a new kind of biological warfare of "specific ethnic genetic attacks." Beijing's relentless efforts to collect genetic profiles of foreigners while preventing the transfer of the profiles of Chinese outside China is an indication of dark military intentions.

Can China actually develop such pathogens?

Technicians can now build a DNA weapon targeting a single individual, so weapons for specific ethnic groups must also be possible. In short, the next disease from China could leave Chinese people immune but sicken and kill everyone else. Call it the Communist Party's "civilization-killer."

Would China spread another disease so soon after COVID-19?

Xi Jinping, the bold Chinese ruler, knows he has just killed more than six million people outside China by deliberately spreading the disease beyond his borders.

The evidence of Chinese responsibility is clear, but the world refuses to admit that the Communist Party could be so malicious. For instance, the American intelligence community, as stated in the unclassified summary of its report delivered to President Biden in August 2021, could not come to a conclusion as to the source of the disease. The intelligence community, the summary stated, "remains divided on the most likely origin of COVID-19."[2]

Biden has remained uncurious about a disease that had

then killed hundreds of thousands of Americans, and he has dropped the matter even though the American toll has now crossed one million. Xi Jinping, therefore, knows he has suffered no penalty for killing Americans in great numbers. As a result, he undoubtedly sees no cost in spreading another disease, especially one that is targeted on the non-Chinese. It does not appear that any deterrence exists.

Analysts have often said that the first moments of a war with China will be fought in low-earth orbit, as both sides move to shoot down or disable the satellites of the other. Perhaps the war starts in outer space, but maybe it begins six months before, when China releases a virus that attacks only foreigners. After all, COVID-19 has been the proof-of-concept of the effectiveness of biological warfare. The disease has crippled – and continues to hobble – societies across the world.

China is a party to the Biological Weapons Convention, which prohibits signatories from developing and maintaining stocks of such weapons. But the Convention includes no inspections regime, which means the treaty will not stop the Communist Party.

The international community, assuming that COVID-19 was a natural occurrence, is still trying to cooperate with China to prevent another pandemic. Instead, countries should realize that the Chinese party-state manufactured and deliberately spread the disease. Governments, therefore, should be taking steps to deter the Chinese regime and building defenses against the next product of China's biological weapons laboratories.

After all, COVID-19 came out of a box, ready not only to infect but also to kill.

The single greatest threat to the United States today is not nuclear weapons or terrorism. It isn't even cyberwarfare or foreign attacks on our democracy. The primary threat America faces today comes from the biotech sector in the People's Republic of China. Only in the biotechnology space could one capture and manipulate the very building blocks of life to destroy one's enemies.

That same technology could also be used to *genetically enhance* the people of the country that the technology is being used in, like in some bad science fiction film. This is particularly true if, as in the case of China, this takes place in an unregulated and opaque system, where so much biotechnology research and development is being done. And let's not even get into (yet) the dangers that improperly regulated experimental biotech R&D pose to the world in the form of major accidents that could fundamentally upend the global order and economy, as the novel coronavirus from Wuhan, China, did in late 2019.

Using a technology called CRISPR, geneticists have the capability to identify, chart, and "edit" the genes that construct our very bodies. Over time, this technology can be developed into a potentially Earth-shattering offensive strategic weapon. This is especially true as the rising People's Republic of China races forward with its national development. More dangerously, the Chinese Communist Party (CCP), which rules China with an iron fist, has no history of respecting the value of the individual, or that of human life in general.

Over the course of this book, you will see how the character of the Chinese regime influences every aspect of its country's development. By understanding that China is one of the most prolific human rights violators in the world, that the Chinese regime is the very embodiment of "might makes right," you will understand just how dangerous allowing this regime to develop the advanced tools needed to dominate the biotech industry is for the rest of the world. The CCP is a regime that has subjugated Evangelical Christians. This is a group that is in the process of colonizing Tibet by sending ethnic Han to the occupied territory in an effort to make the Tibetans a minority in their own land. The CCP is currently in the process of ethnically cleansing the Muslim Uighurs from the Xinjiang Province.

Meanwhile, this regime systematically represses the democratic people of Hong Kong.[1] Dotting China is a vast network of slave-labor camps – known as *Laogai* prisons – wherein political prisoners are made to work.[2] Falun Gong prisoners are routinely executed on trumped up charges so that Chinese authorities can harvest their organs and sell them on the black market.[3] The CCP uses its immense wealth and power quite literally to spy on every single one of its citizens with nationwide closed-circuit camera systems.[4] It denies its people the ability to access all of the worldwide web, out of fear that "foreign ideas" will foment a revolt against the CCP's totalitarian rule.[5]

Until 2014, also, there was a decades-long state-enforced "one-child" policy that led to the infanticide of millions of unborn children. Needless to say, this is a regime that should not be trusted with the power it already wields. Allowing the CCP to conduct untrammeled biotechnology research and development is only asking for trouble.

The biotech battle between China and the United States is not merely confined to the realm of scientific research. It crosses a multitude of fields and plays out in unpredictable, downright dangerous ways. For instance, the biotech battle is fought in the arena of intellectual property (IP) law. China is a persistent and flagrant violator of international IP laws.

China is a country that one in five US corporations accuse of having stolen their intellectual property at some point in the last decade. According to the 2017 report from the Commission on the Theft of American Intellectual Property, "Chinese theft of American IP costs anywhere between $250 billion to $600 billion annually."[6] In every sector, from manufacturing to computers to biotech, China has stolen a treasure trove of proprietary data, all in an effort to compete with the Americans and to leapfrog US companies in the great race for world technological dominance.

China's biotech battle with the United States also plays out in the international business domain. Not only have Chinese spies stolen IP from American companies, but, paradoxically, American businesses and entrepreneurs have entered into partnerships with Chinese firms. Even as the US and Chinese governments tussle with each other at the geopolitical level, the business and academic communities of the two rivals cooperate with each other on a routine basis. What's more, as you will see, the US government assists the Chinese government in sensitive biotech R&D projects!

During the Cold War, America's private sector was very often the silver bullet to overcoming the centrally planned leviathan of the Soviet Union. In today's new cold war, America's private sector routinely cooperates with China. In their exchanges, Chinese firms (and therefore the Chinese government) learn about innovative new business practices and proprietary research from Americans that these Chinese firms then incorporate into their own products. With knowledge gleaned from their American counterparts, the Chinese firms then create products that are meant to empower China at the expense of the United States.

China has targeted American academia for exploitation. Given that biotechnology is part of the "knowledge-based economy," academic research plays a large role in developing new products for this industry to sell. In recent days, there have been various reports of extensive Chinese efforts to steal proprietary research from American and Canadian firms or publicly funded research institutes.

In some cases, as you will see in future chapters, American research institutions have happily shared information with their Chinese counterparts, only to then be undercut by those same Chinese labs. In fact, several Chinese researchers have been sent to the United States to receive US taxpayer–funded research grants from the National Institutes of Health (NIH). Once they receive NIH grants, the Chinese scientists then gain access to proprietary research data, steal that data, and send it back to their labs in China.[7] Since 2015, Chinese industrial espionage directed against the biotech sector has been concentrated on stealing data regarding coronavirus, cancer, HIV, and Ebola research.

More dangerously, China has successfully targeted American researchers, offering those researchers money and prestige in China (far more than what they often get in the West), resulting in a reverse brain-drain of talent and knowledge bleeding from the United States into China.[8] All of this plays well into China's attempts to create an indigenous base for high-end biotechnology research and development. As its indigenous biotech capacity increases, so too does interest from foreign venture capitalist firms and foreign biotech firms.

China's autocratic president, Xi Jinping, identified biotechnology as one of seven major tech industries that China had to dominate in order to control the battlefield of the twenty-first century. Chinese generals have written extensively over the last few years about the offensive capabilities that CRISPR would give their forces. Notably, China's military leaders have highlighted how biotechnology could allow China's forces to conduct "specific ethnic genetic attacks" on individuals and groups with whom they disagree.

Other Chinese militarists have spoken about how the country that rises to dominate biotechnology first will have exclusive access to essential "biomaterials" and even "brain control weapons."[9] What's more, under President Xi's rule, the entirety of the Chinese biotechnology sector has been reorganized to allow for the People's Liberation Army (PLA), the military arm of the totalitarian CCP, to enjoy greater access to and control of China's budding biotech sector.[10] President Xi is believed to have authorized the funding of direct Chinese military biotechnology development on the tune of $1 billion per year. And the PLA has overseen experiments using CRISPR to augment the genes of select soldiers among their ranks – a process known as "gene doping" – in an attempt to create genetically enhanced super-soldiers.

In fact, there were concerns for the security of the DNA of athletes participating in the 2022 Winter Olympics in Beijing.[11] Fears abounded that China would demand samples from these top physical specimens, as part of its COVID prevention protocols and use that as a backdoor way to gain access to desirable genetic traits that it could conceivably use in its military gene-doping program. This might sound ludicrous, but gene-doping – hacking human biology for strategic or economic gain – is the wave of the future, and the Chinese are at the bleeding edge.

Unlike here in the United States, where there is a degree of separation between our military and the scientific sector, the state and indigenous private industry in China have fused. Yes, there are technically "private" corporations and research institutes in China. But these groups are almost always staffed by or receive funding from Chinese military and Communist Party groups. The purportedly "private" biotechnology sector in China has exploded. Whereas five years ago, biotech research and development in China received funding from venture capital (VC) of around $1 billion per year, today China's biotech sector receives VC investment on the order of $12 billion per year.[12]

That figure is expected to increase as both interest and capabilities in China's biotech sector increase. This is especially true in the wake of the COVID-19 pandemic. Because of the pandemic and the race to find a cure, biotech has secured itself as a massive growth industry that is highly attractive to investors.

Also, unlike in the United States, the Chinese biotech sector is a relatively unregulated space. Experiments involving the successful cloning of monkeys, dogs, and horses have occurred in China over the last few years.[13] Chinese researchers have also managed to create "chimera" animals, such as pig-monkey hybrids.[14] All of these actions have been taken to further the capability for science eventually to clone both human organs and, potentially, entire human beings.

Many medical researchers in the West applaud – and even

assist – China in these activities. Since the biotech industry in the United States is regulated far more than its counterpart in China, Western scientists, investors, and entrepreneurs are flocking to China's growing biotech sector *because of its deregulated environment.* In this environment of freewheeling deregulation, American innovators can work in tandem with Chinese researchers and firms to develop cancer vaccines, Ebola treatments, and HIV drugs. They also conduct unchecked genetic experimentation with CRISPR while yearning for the day when they can successfully clone human organs to be used on the international organ transplant market.

All of these things could unequivocally enhance human health everywhere. Given that the research is being conducted alongside Chinese state entities, however, you can be certain that there will be many more negative impacts on human existence. After all, as the Chinese military has made clear, the biotech industry offers significant strategic offensive capabilities to whichever country captures it first.

As I have written elsewhere, China has embraced the *Field of Dreams* mentality. If China's government builds the infrastructure needed to conduct high-tech R&D in China, the foreign researchers and investors will come.[15] China has spent years developing the tools needed to seduce ambitious, ignorant, and greedy Westerners to do business in China.

We've seen this pattern play out in the manufacturing sector, we see it presently at play in the computer industry, and we are starting to see it in progress in the biotech field. Through its "Thousand Talents" program, the CCP has a concerted, well-funded project underway to lure America's best and brightest to come to the competitive, freewheeling, leap-without-looking high-tech R&D sectors of China.[16] The Chinese have targeted American doctoral students enrolled in the nation's top universities for co-option; they've wooed noted researchers such as Harvard's Charles Lieber, who was recently convicted by the Department of Justice for lying to the FBI about his close financial ties to Chinese research labs. The list goes on.[17]

China is awash in cash, and its spies have no problem throwing money around to win the cooperation and trust of unwitting (or immoral) American researchers. As the Chinese artificial intelligence researcher and high-tech venture capitalist Kai-Fu Lee wrote recently of Chinese entrepreneurs,

> They live in a world where speed is essential, copying is an accepted practice, and competitors will stop at nothing to win a new market. Every day spent in China's startup scene is a trial by fire, like a day spent as a gladiator in the Coliseum. The battles are life or death, and your opponents have no scruples. . . . The only way to survive this battle is to constantly improve one's product but also to innovate your business model and build a "moat" around your company. If one's only edge is a single novel idea, that idea will invariably be copied, your key employees will be poached, and you'll be driven out of business by VC-subsidized competitors. . . . The messy markets and dirty tricks of China's "copycat" era produced some questionable companies, but they also incubated a generation of the world's most nimble, savvy, and nose-to-the-grindstone entrepreneurs.[18]

So, you see, the copying of intellectual property and the race to the top of the tech industry is not merely something the Chinese do against the Americans. It starts much closer to home, in China's domestic market, and proliferates out from there. How can we expect such a rambunctious – *rapacious* – Chinese industry and government to treat our firms and intellectual property any different than they treat themselves? Apply Kai Fu-Lee's observation specifically to the biotech sector and one should be put on the defensive. China has our number.

As a geopolitical analyst specializing in the interaction of new technologies with international relations (what some have come to call "geotechnology analysis"), I've spent years studying biotechnology research and its national security implications. What's more, I am married to a nurse scientist who previously worked in genetics at the NIH. Biotechnology has been an ever-present factor in my professional and personal life. In fact, since 2018, I've covered the biotech beat for *American Greatness*. I was warning about the threat that China's ambitious biotech program posed to the world well before the COVID-19 pandemic.

With CRISPR-Cas-9, Chinese geneticists can gene edit whatever they want. In one instance, a pair of HIV–infected twin girls were the subject of gene editing experiments while they were still in their mother's womb. The object of the experiment was to remove a specific gene in the twins' DNA that would prevent them from being infected with HIV.

By conducting the tests *in utero*, the Chinese scientist who conducted the experiment, He Jiankui of the Southern University of Science and Technology in Shenzhen, China, ensured that the twins' relative immunity to HIV would be passed on to their children (the·twins' parents were HIV positive).[19] The children were born healthy in 2018. Subsequent tests on the children have indicated that the specific gene that made them susceptible to HIV was removed, but those gene edits caused further mutations in their DNA. In fact, the children have shown unspecified enhancements in their brains specifically because of the experiment they were subjected to while still *in utero*.[20]

When the news of these experiments went public, despite their success, the Chinese government charged the scientist who conducted them, He Jiankui. Dr. He was arrested and found guilty by a court in Shenzhen of having conducted illegal biomedical practices. He was fined three million yuan ($429,000) and sentenced to three years in prison in 2019. Since his sentencing, He's whereabouts are unknown. Beyond that, He's two research partners who partook in the experiments were also punished by China's authorities.[21]

Tellingly, the Chinese government has not revealed the details of the investigation into He and his experiments. While Beijing was clearly concerned about free-wheeling biotech experimentation being undertaken by individual scientists in China, it must be noted that China's government has not reined in the wildcatting experimentation in which most Chinese scientists and biotech firms are currently engaged. It is likely that China's leaders did not want the controversy in the international press; that Beijing was worried negative press would harm the overall biotech sector in China, which China's leaders believe is an essential aspect of China's comprehensive national strategy for achieving global dominance. So, He Jiankui was sacrificed on the altar of international public opinion to shield China's other, riskier biotech endeavors from greater international scrutiny.

He Jiankui and others who have engaged in such incredible, even reckless, biotech experiments claim they did what they did to enhance and protect human life. Even if the Chinese scientists conducting these fly-by-the-seat-of-your-pants biotech experiments on animals and children alike are well-intentioned, the fact that they are conducting these sensitive experiments in such a lax regulatory environment indicates that something disastrous could go wrong – and that the world would not know it until it was too late. Think about He Jiankui: here is a man who conducted unproven experiments on babies still in the womb. His stated intention was to prevent them from being infected with HIV as their parents were.

But He Jiankui and his team also enhanced the brains of these children.

If this was done purposely (a possibility), that would be frightening enough. If this was an unintentional outcome, that should still worry you. It is believed that untrammeled gene editing could cause unintended changes to DNA. These changes might not be noticed until after the initial gene edits were done with CRISPR-Cas-9 and could create more problems than they solve. In essence, the Chinese are playing with technology they have little understanding of – much of which was either purchased or stolen from the West – all so that they can beat the Americans. Thus, the Chinese government's support and encouragement of risky biotech research and development is *not* scientific understanding, as is usually the case for the West. Instead, their intentions are purely geopolitical, as Gordon G. Chang alludes to in the foreword for this book.

With the recent outbreak of COVID-19 and its subsequent transmission to the rest of the world, some have questioned its origins in Wuhan. The fact that Wuhan is the center of much research on coronaviruses cannot be overlooked.[22] While most Western scientists have dismissed the very notion that the disease was created in the Wuhan virology lab, some prominent

Americans – notably Senator Tom Cotton (R-AR) – insist that there is more to this story.[23]

My colleague at *American Greatness* Steven W. Mosher has rightly demanded that China release all relevant data sets from its virology lab in Wuhan, where many believe the coronavirus may have originated.[24] There are some who claim that this was a bioweapons attack. Given the slapdash nature of biotech research and development in the People's Republic, it's far more likely that the lab conducting the research on the novel coronavirus lost containment of COVID-19 and accidentally loosed this plague upon the world. Still, later in this book I will walk through why COVID-19 may very well have been a bioweapons attack, targeted at the heart of our democracy: the 2020 presidential election. After all, COVID-19 served China's strategic interests well. The United States endured a political regime change in the aftermath of the pandemic, its society has been irrevocably altered (embracing China's draconian approach to disease mitigation and social controls), and its economy has been fundamentally weakened. Specifically, the pandemic and its aftermath precipitated the removal from power of America's unpredictable forty-fifth president, Donald J. Trump, who was making himself and the United States a direct threat to China during his tenure. Had it not been for the pandemic, it is likely that Trump would not have lost reelection and the United States would still be the vibrant, free nation it was in 2019. Alas, we are made to endure a "new normal."

This could be why the CCP covered up the outbreak in Wuhan from the world. In fact, the pathogen appeared in Wuhan far earlier than the CCP officially acknowledged. Senator Cotton and others claim that China hid the outbreak from the world not only because it would embarrass the regime, but because it was attempting to hide any evidence that COVID-19 was, in fact, from its lab.[25] In so doing, at the very least, the Chinese put the rest of the world at risk.

American research institutions associated with the NIH have ties to the virology lab in Wuhan. The University of North Carolina Chapel Hill has extensive scientific partnerships with Chinese labs.[26] So does the University of Maryland. They are not alone. As you've seen above, the Chinese have not only openly partnered with American academic research programs, they've also striven to pilfer research data from NIH–funded projects specifically related to coronavirus, Ebola, HIV, and cancer.[27] Incidentally, the COVID-19 strain has elements of Ebola, HIV, and cancer embedded within its genetic makeup (this explains why some Ebola and HIV treatments have been successful in also treating cases of COVID-19).[28]

The scientific community insists that the presence of these genetic markers is simply an indication that it came from nature; everything on Earth is, after all, from the tree of life. But one cannot help but to wonder *why* the Chinese were so interested in obtaining research related to these specific diseases from the US. The COVID-19 strain is a chimera just like those monkey-pig hybrids that Chinese geneticists tinkered with a couple of years back. In future chapters, I will more deeply explore the controversy surrounding the origins of the current pandemic and endeavor to clarify whether there is truth behind the more salacious claims made by Senator Cotton and his ilk, or if COVID-19 truly is a natural disease.

One thing is certain: Americans everywhere should be more aware of the kinds of experiments being done in China's biotech sector (and to a lesser degree, the abuses our own biotech sector is engaged in). Those readers who are not in the biotech or national security sectors are likely unaware of the immense progress China has made in becoming a potent biotech power. Perhaps there are even professionals in both American biotech and national security who are unaware of these trends. Do not rest on the assumption, simply because the United States has long been the source of high-tech innovation, that it will always remain so. This was precisely the view that Britons took of themselves not long after the United States separated from the British Empire.

As I described in my first book, *Winning Space: How America Remains a Superpower* (2020), the upstart, agrarian Americans were able to compete with their British cousins in the nascent industrial sectors early on because of what today might be described as "industrial espionage" with a touch of "intellectual property theft." In 1789, a British citizen, Samuel Slater, who had spent his youth working in the advanced textile mills of England, emigrated to Rhode Island and essentially created a modern British factory on America's East Coast. According to a *PBS* biography on Slater:

> He dreamed of making a fortune by helping to build a textile industry. He did so covertly; British law forbade textile workers to share technological information or to leave the country. Slater set foot in New York in late 1789, having memorized the details of Britain's innovative machines.[29]

Using his knowledge of British textile practices, Slater is remembered as the "father of the American factory system." Ultimately, Slater created an entire factory town, known as Slatersville, in Rhode Island. "Slater divided factory work into such simple steps that children aged four to ten could do it – and did. While such child labor is anathema today, American children were traditionally put to work around the farm as soon as they could walk, and Slater's family system proved popular." Beyond that, Slater's "Rhode Island System" became popular with other budding American industrialists seeking to compete against the dominant British textile industry. Great industrialists like Francis Cabot Lowell would eventually imitate and expand upon Slater's factory system.

In England, Samuel Slater is not remembered fondly. Most Britons remember him by the epithet of "Slater the Traitor."[30] Today, countless American entrepreneurs, innovators, and investors – to say nothing of large firms – have done business in China's high-tech sector. To gain access to China's vast market share and its dynamic, sophisticated high-tech research and development infrastructure, these American companies, researchers, and labs must willingly share proprietary information and methods with Chinese state firms, or else be prevented from doing business in the country. The Americans engaging China in high-tech research and development are doing for

China what Samuel Slater did for the young United States in its industrial competition with the British Empire.

Without Slater's input it is likely that the Americans would not have managed to keep up with British industrialists – let alone leapfrog them by the twentieth century. At the time, the British wrote off America's rising threat to their industrial dominance, just as Americans today downplay the Chinese threat to the United States' dominance in advanced technology. We do so at great risk. In fact, China has managed to catch up with America on multiple fronts in what Gordon G. Chang has rightly described as "The Great US–China Tech War."[31] Now, China appears poised to leapfrog the United States in critical high-technology areas such as biotech.[32]

China behaves very much as the United States of old. Through guile, ruthlessness, and innovation, the Americans were ultimately able to outmaneuver and best their British cousins, achieving global dominance. Similar strategies are at play in China today. But the grave risk from untrammeled Chinese biotech research and development must be more fully understood.

Should such research continue in China – and should China grow to become the leading biotech power – the Chinese could have the capability to launch "specific ethnic genetic attacks" against their American rivals.[33] Or, they could deploy gene-doped super-soldiers onto the battlefield against their enemies.[34] Chinese biotech researchers have already become enamored with the prospect of marrying their research with China's prolific artificial intelligence research program.[35]

Since 2016, the CCP has run the National GeneBank, probably the world's largest bioinformatics program at present.[36] The CCP has collected massive stores of data from Chinese citizens, and it plans on using artificial intelligence to sift through the bioinformatics to help Chinese scientists design "therapeutics or *enhancement* [emphasis added]."[37] Already, the CCP has forcibly collected genetic data from Muslim Uighurs it has imprisoned in Xinjiang.[38]

There is real concern on the part of human rights activists

that the constantly evolving biotech sector in China could soon use its immense capabilities to control or selectively eliminate entire ethnic groups.[29] Further, Beijing Genomics Institute (BGI) has become a front group for the CCP's bioinformatics revolution. This company has partnered with American biotech firms in California, as well as the Children's Hospital in Philadelphia.[40]

Some national security officials believe that BGI has unprecedented access to the personal genetic information of countless Americans, in much the same way that Beijing gains access to the personal financial information of Americans through cyberattacks.[41] And just like Chinese hackers do with the personal information of Americans stolen through cyberattacks, Chinese firms could wantonly sell such information on the budding bioinformatics black market.[42] Further, Chinese firms could engineer diseases specific to certain American groups – possibly even individuals. Or these Chinese firms could simply blackmail Americans by possessing this data.

At the very least, COVID-19's emergence in Wuhan should have highlighted the grave danger that freewheeling experimentation in China's biotech sector poses the world. Therefore, in the last sections of this work, I intend to advocate for the formulation of international legal norms governing the use and conduct of biotech research. This will require the United States to lead. It will also require the US government to work alongside its partners to deny China the privilege of gaining access to proprietary Western biotech research.

Time is not on our side. As China's biotech sector continues to evolve, its ability to use biotech as a strategic weapon will only increase. Unless American policymakers can recognize the depth of the threat and stop China now, there may be no keeping it from making the world sick, either intentionally or accidentally.

And, like Pandora's Box, once the demons are released, there may be no return. Chinese biotechnology dominance could very well mean the beginning of – to use a tired phrase – a "new normal" wherein China leapfrogs the West by using devastating and debilitating biotechnology attacks to force lockdowns, spread illness to specific groups of people, and sow fear and discord within America's national polity. The 2020s, like the 1910s, could very well see the outbreak of a world war that unleashes pathogens and plagues the likes of which fundamentally change the human race itself.

As it stands, the United States is entirely unprepared for this possibility – as the slipshod response to COVID-19 proved. Interestingly, the only major economy to see growth during the 2020–21 period was that of China, the place from whence the terrible plague came.[43] Meanwhile, the pandemic utterly disrupted life on all levels in the West, specifically the United States. The pandemic likely caused a major political regime shift in America's 2020 presidential election. As the economy collapsed under the pressure of national lockdowns, they understandably blamed President Donald Trump and opted instead to vote for his Democratic Party rival, Joe Biden.[44] As it happens, Trump may have been better suited for dealing with the Chinese threat than Biden ever could be. But the fog of COVID-19 clouded everything, and it is only just now lifting.

While the economy in the West has begun a tepid recovery, most experts believe that it will take years for it to return to its 2019 peak. This is to say nothing of the logistical crises caused by COVID-19 and the subsequent lockdowns. American leaders, Democratic and Republican alike, must take note of COVID-19's damage; they must be willing to understand and appreciate the threat that untrammeled Chinese biotech research and development pose; and they must be willing to take necessary – even drastic – steps to ensure that another biological crisis, one that is potentially more dangerous than COVID-19, is avoided at all costs.

CHAPTER 1
WARFARE THROUGH OTHER MEANS

Since the Second World War, Mankind has possessed the tool of its self-destruction. Since humanity mastered the atom and developed atomic weapons, the world has danced on a radioactive knife's edge. During the Cold War, the United States and the Soviet Union were brought to the brink of nuclear annihilation a few different times. Again, as the Russian Federation has waged war on its neighbor, the proto-democracy of Ukraine, fears abound that nuclear war between the West and Russia is at hand.

But there's a new actor vying for the starring role in humanity's ongoing psychodrama: biological warfare. Having long been cast in a supporting role in the psychodrama, today, it has quietly won the affections of multiple producers – notably the Chinese Communist Party. Since the eruption of COVID-19, biowarfare has become the new "it girl" of human conflict. And like every diva, she is going to jealously guard her newfound stardom.

Since the Cold War, the world's great military powers have maintained stocks of biological agents that could be used to kill millions of people in rival nations. But the great military powers recognized that biological agents evolve and act in unanticipated ways. Unlike nuclear weapons, which simply kill everyone in a targeted area, the microscopic nature of bioweapons is such that there's no guarantee enemies and allies alike won't be killed in the attack. Historically, these military powers – notably democracies, like the United States – have looked upon bio-weapons with a degree of hesitation.

But totalitarian powers, seeking leverage and advantage over their rivals, have long sought out a "silver bullet" with which to destroy their enemies with one fell swoop. Such powers, like the old Soviet Union, Saddam's Iraq, North Korea, or today's People's Republic of China, care little about human rights or the laws of war. In fact, these powers have often derided the so-called "laws of war" as mere constructs of Western policymakers meant to perpetuate Western dominance. These totalitarian states have striven to gain technological supremacy over their demo-cratic rivals and have long feared direct military confrontation with Western armies in the modern age. Rather than fight on terms the Americans and their democratic allies have preferred, the world's totalitarian powers have favored fighting unconven-tionally, living up to Sun Tzu's dictum of "winning without fight-ing." Or, at the very least, winning by fighting indirectly.

China has mastered the art of indirect, unconventional warfare. Beginning in the 1970s, China's rulers used trade as a weapon with which to raid critical industries in the United States. American elites misinterpreted trade with China as an economic policy decision, but for Beijing it was war by other means. Sapping the United States of its prodigious manufactur-ing and productive capability allowed for China's Communist Party to build the world's most robust manufacturing sector in the world. China effectively became the sweatshop of the world, and every major Western nation came to rely upon the cheap goods built in China. Yet what most Westerners considered to be slave wages at the time, many Chinese workers were grateful for having. In the context of China's economy, the small amount of money they were paid was enough to allow for these other-wise impoverished and oppressed masses to build economic clout for themselves and their families. In turn, after many years of serving as the world's sweatshop, China built the world's largest (and growing) middle class.

The "trade war" that China waged worked brilliantly. As Luigi Zingales illustrates in his 2005 book *Capitalism for the*

People, China was not finished when it became the largest blue-collar manufacturing economy in the world. It wanted dominance in the budding knowledge-based global economy. China next came for America's white-collar jobs; it wanted America's high-tech industry. Since the 1990s, China has increased its share of highly skilled, white-collar workers; it has created the world's largest base of well-trained scientists and engineers, too. After thirty years, this has had real-world implications for the dominant position that American leaders assumed they'd simply always possess.[1]

In the 2020s, China is now ready to challenge the American-led world order that has persisted since 1945. Yes, the Americans still possess the deadliest military in the world. But the Chinese do not seek to fight the Americans on an even footing. Since the Napoleonic era, war has become increasingly democratic. Before the Napoleonic Wars, during which France and other major powers began conscripting soldiers and sailors from the masses of the people, European wars were waged by a cadre of elite soldiers. There were strict, almost gentlemanly rules agreed upon by all combatants. War was more akin to a duel. Today, it is not only the masses of ordinary people who are called to war; it is entire societies. There is no longer a "home front" and a battlefront. The lines are blurred. The battlefront is the home front and vice versa; at least, this is how the Chinese see things.

Before the twenty-first century, biological warfare was taboo. The Soviets were the most vicious experimenters with biological warfare agents and theories. When the USSR vanished into history's good night, most assumed no other power would ever challenge the American juggernaut that was riding high from its Cold War victory. Sadly, warfare is not final, and America still had implacable foes, looking to exploit whatever weaknesses they could locate. The American security establishment was highly attuned to the threat that nuclear warfare posed, despite the Cold War having ended. So, China and others crafted unorthodox weapons with which to sap America's martial strength. Cyberwarfare, counterspace warfare, chemical warfare, and biological warfare all became areas of interest.

Biological warfare has especially evolved. Historically,

biological warfare and the agents used in the development of bioweapons were easily identifiable. They could also be regulated and monitored by the world's powers. The advent of sophisticated biotechnology has changed this calculus, making biowarfare more likely. This is due to the globalization of biotech research and development, thanks to arrogant American scientists who are more interested in "trusting the science" than they are in protecting the national interest, as well as to ignorant government officials who failed to understand how China was penetrating America's most secretive institutions and stealing our proprietary biotech trade secrets (more on that later). China today has become the world's leader in biosciences. This represents a threat that the United States has not encountered since at least the Second World War. Washington faces a foe in Beijing that is, in some cases, a technological equal – and in some key places, a technological superior.

If power abhors a vacuum, then international relations resent weakness. Sensing that America's Comprehensive National Power (CNP) is waning and, at least for now, China's is on the rise, Beijing will not kowtow any longer to the United States. The time is at hand for China to move the geopolitical map around in its favor. Biowarfare offers one of many unconventional ways for China to accomplish this task.

Looking out from China, President Xi Jinping sees what he believes to be a "breakaway province" in democratic Taiwan. Just across from what's known as the Taiwan Strait, an island of almost 24 million people live in freedom. The island itself has been viewed as an "unsinkable aircraft carrier." Given its democratic inclinations, Taiwan has long been a recipient of US military support and diplomatic friendship. Many Chinese, especially the leaders in Beijing, view Taiwan in much the same way as Abraham Lincoln looked upon the American South in the Civil War: a rebellious province, not an independent nation. But unlike the American South, Taiwan has developed independently from the Communist Chinese system for decades; it has struggled for and maintained its freedom, despite Beijing's best efforts. The Taiwanese even speak their own unique dialect of Mandarin.

Yet the strategic ambitions of China's rulers, to push US

military power out of the Indo-Pacific, which Beijing believes to be its exclusive sphere of influence, is profound. Taiwan is key to Beijing's grand strategy of regional (and eventually world) domination. Today, China has the resources and military capabilities to accomplish the task of invading and conquering the small island. There's little belief that the Americans, despite possessing an alliance with Taiwan in the form of the Taiwan Relations Act, will risk a major – even nuclear – war with China to stop their invasion of the island democracy.

Still, every analyst who concedes that China has the power to invade Taiwan today believes it would be a brutal, bloody war that would extract a heavy price for China – notably, its ruling class. The Pentagon does not believe China will attempt to invade Taiwan until closer to 2027, if ever. But few in the US national security establishment appear to understand the role that total unconventional warfare plays in the minds of China's strategists. The first thing that China will do is attack the Americans and their allies in unanticipated ways. Hence, biowarfare.

Consider the COVID-19 pandemic as a proof of concept. China learned that the Americans may be all-powerful militarily, but when it comes to germs, their purportedly free and strong civil society falls to pieces. The COVID-19 pandemic was the ultimate societal stress test, and the Americans (and much of the West) failed. Sure, China got knocked around hard by the pandemic, but, as you'll read, it came out of the crisis still growing economically, still enhancing its war machine. The Americans, meanwhile, are a mess. And the COVID-19 pathogen was by no means the most dangerous in history. China now understands that it can attack the civilian struts of our power, which, in turn, would undermine our ability and will to facilitate an effective defense of Taiwan, or that of any other regional ally targeted by China.

Biowarfare today is not your father's biowarfare. It is not a distinct area of investigation restricted to the military sciences. Instead, it is entirely dual-use. In many cases, as you'll read, unwitting Western scientists (and maybe even some fellow travelers for China's Communist Party in the West) have played a critical role in enhancing China's indigenous biotechnology sector. In turn, this indigenous biotech sector has been managed and supported by the Chinese People's Liberation Army. In fact, it is likely that the entirety of China's supposedly civilian biotech sector is actually a front for the PLA's expansive bioweapons program that, even now, is being honed to launch the mother of all biological first strikes on Americans and other enemies of China.

As Donald Rumsfeld loved to say, "We are on notice, but we have not noticed." China has cracked the code for beating the Americans. It has figured out that most American leaders and people are unable to see warfare as anything other than a binary affair: we either are or are not at war. For Red China, war is a spectrum of grays, and China is waging war in countless ways. China behaves as the Lilliputians in *Gulliver's Travels*, tying down the gentle giant with unanticipated swarming tactics. Biowarfare in the form of pandemics aimed at the civilian population and economy is the best way for China to stymie the American military juggernaut before it can even mount an active defense.

China would love nothing more than to remove the US military from the Indo-Pacific without ever having to fight us directly. A pandemic would do the trick, as COVID-19 proved. It'd drive us nuts. With the added fuel of disinformation, an offensive biological attack in the form of a supposedly natural pandemic would turn Americans on each other. The United States would either collapse or simply turn inward and ignore the real threat of the CCP. All that military might would mean nothing if the political and economic system were collapsing. This is how China's rulers think. This is the strategy they prefer. And, thus far, the Americans have proven to them that the biowarfare domain is the true frontier of warfare. If Beijing dominates this area, it controls the building blocks of life itself.

In fact, given how badly the Americans and their allies responded to the pandemic on a societal level – at how blinded we all were to the fact that COVID-19, if not a direct biological weapons attack then likely the result of a lab leak that China's rulers promptly covered up with the help of American scientists and leaders – China could first launch a silent biological attack on the United States and then offer us aid in our hour of need, ensnaring us even further in its web of lies. For China, it seeks to restore its historical position in the world as the "Middle Kingdom." It wants to be the center of the world system. And it wants to make all other states – including America – nothing more than vassals paying tribute to its supreme power. If warfare is nothing more than politics through other means, as Clausewitz says, then China has mastered this art of war far better than the Americans have.

CHAPTER 2
LOOSE LIPS (AND COVID-19) SINK SHIPS

The U.S.S. *Theodore Roosevelt* (CVN-71) is one of eleven *Nimitz*-class nuclear-powered aircraft "supercarriers" in the United States Navy.[1] With a height equal to that of a twenty-four-story building, carrying about five thousand crewmembers, and maintaining around eighty different aircraft, the *Roosevelt* is the equivalent of a floating town.[2] The *Theodore Roosevelt* cost around $4.5 billion to build.[3] The supercarrier and its attendant battle group costs $6.5 million *per day* to operate. Because the ship is powered by two Westinghouse nuclear reactors – and given its size and the large number of combat aircraft the ship carries – the modern US Navy aircraft carrier is the ultimate symbol of American power projection. As a popular internet meme from the early 2000s says, American aircraft carriers represent "100,000 tons of diplomacy."[4] These warships are not only the backbone of the modern US Navy, but also the proverbial tip of America's spear.

America's aircraft carriers are often the first line of defense, the initial point of contact between American military power and the rest of the world. Whether it be responding to Chinese aggression in the South China Sea, reinforcing Taiwan, or deterring North Korean nuclear brinkmanship, the US Navy relies on its flattops. Even for humanitarian aid missions, aircraft carriers are often deployed to lead rescue and recovery efforts because they are such large vessels with reliable aircraft.

It's like having a mobile, floating military base capable of going to any place in the world.

The United States has operated aircraft carriers since before the outbreak of the Second World War. Back then, the Navy's war planners assumed that the battleship would be the primary weapon used in war. But when the Japanese Empire conducted its devastating surprise attack on Pearl Harbor and knocked out most of the battleships in America's Pacific Fleet, what Americans had left at their disposal were aircraft carriers. And the aircraft carrier ended up winning the naval war in the Pacific Theater.

No American aircraft carrier has been lost in combat since World War II. Today, we think of them as invulnerable. They are not. In fact, these great ships are only as strong as their weakest link. And very often, the weakest link is the human element. Maintaining a crew of five thousand sailors and marines on the edges of the world for months at a time is a logistical feat in itself. And having a crew that large allows for redundancy, meaning the ship can still be effective in combat even if it loses some crew in fighting.

But what happens when the crew falls sick? Or, more worryingly, what happens when an American aircraft carrier is attacked with biological weapons?

Apparently, no one at the Pentagon thought about what might happen if a disease spread throughout the crew of one of its supercarriers. In 2020, as the novel coronavirus propagated from Wuhan, China, to the rest of the world, the crew of the *Theodore Roosevelt* was exposed to COVID-19. And the *TR* became a glaring example of how unprepared the Navy's best ships were for a biological surprise attack.

After departing for a routine deployment from San Diego to the Western Pacific Ocean in January 2020, the great ship eventually arrived in Da Nang, Vietnam. As per the *US Naval Institute*'s report, on March 5, some members of the crew were involved in a four-hundred-person reception at a hotel in Da Nang.[5] It is believed that those in attendance were unknowingly exposed to the novel coronavirus. Two weeks after the port stop in Da Nang, the crew of the *Theodore Roosevelt* began exhibiting symptoms *en masse*, including the ship's commanding officer, Captain Brett Crozier.

The pandemic outbreak aboard the *Theodore Roosevelt*, which peaked on March 30, infected 25 percent of the ship's crew. The pathology of the coronavirus indicated that there was an exponential growth pattern in the illness for affected populations. This proved to be true aboard the *TR*. Soon, the whole ship was impacted by the disease. Operating an aircraft carrier requires a crew operating at peak efficiency. An exponentially growing on-board pandemic will nullify any combat effectiveness and render the warship useless, no matter how many redundancies were initially in place.

One could be forgiven for assuming that a warship, with its airtight bulkheads and self-contained spaces, might be an ideal place to quarantine a virulent illness like COVID-19. The *Teddy Roosevelt*, however, was the worst possible place to get sick. The ship's sailors were packed into the warship like sardines in a can. So, no matter how conscientious the crew was in implementing "Bleach-a-Paloozas," the process of scouring the ship routinely with bleach, distributing hand sanitizer to all crewmembers, and instituting strict handwashing protocols, there was no stopping the spread of the respiratory illness. Within a short amount of time, the *TR* was rendered combat ineffective by the disease. Unable to contain the spread of the disease onboard, and unsure of how to proceed, the Trump Administration opted to dock the ship at the port in Guam.

The ship's skipper, Captain Brett Crozier, believed his ship and crew were in immediate danger. He drafted an ambitious plan to evacuate his stricken ship and have his 4,800 crewmembers quarantine on Guam. On March 26, Crozier submitted his evacuation plan to his commanding officers. After four days of what Crozier believed was dithering on the part of his commanders, he eventually submitted his evacuation plan to a wider circle of colleagues – including fellow captains who were outside of his immediate chain-of-command. Because Crozier circulated his evacuation plan outside of his chain of command, one of his colleagues leaked the document to the press. Ultimately, the *San Francisco Chronicle* published an unredacted version of the letter.

LOOSE LIPS (AND COVID-19) SINK SHIPS

Crozier created a needless debacle during a major crisis – the pandemic – by sending his evacuation plan outside of his chain of command. In the letter that was republished by Western media, Crozier is quoted as writing that, "We are not at war. Sailors do not need to die. If we do not act now, we are failing to properly take care of our most trusted asset – our sailors."

Captain Crozier was in an unenviable position. As the commanding officer, the captain was responsible for not only the multi-billion-dollar warship under his command, but also the men and women who served aboard her. These are people he works with and lives among for months out of the year. Anyone can understand where the captain was coming from.

Yet Captain Crozier was not only charged with the safety and security of those under his command, as important as they are. He was responsible for safeguarding one of the most lethal and expensive warships in the history of naval warfare, for ensuring the operational security and integrity of the *Theodore Roosevelt*. Part of maintaining the operational security of the carrier was to hide the true capabilities and vulnerabilities of his warship – especially as it was in-theater.

Sun Tzu teaches that deception is a key tenet of the art of war. During the Second World War, a popular American saying was, "Loose lips sink ships." Similarly, Winston Churchill's quipped to his subordinates that "In wartime, truth is so precious that she should always be attended by a bodyguard of lies."[6] China's former leader, Deng Xiaoping, advised future Chinese leaders to always hide their true capabilities and bide their time.[7]

Captain Crozier believed that the United States was not at war. Technically, he was correct. But the United States has increasingly found itself besieged by rivals – not only China, but also Russia, Iran, and North Korea. In the modern era, as Sean McFate has said, the American view of warfare is akin to pregnancy: you either are pregnant or you are not.[8] Thus, Americans have a binary view of war and peace. Increasingly, however, America's foes do not.

Many American rivals – notably, China – view war as existing on a spectrum. The Israelis refer to this as the "Gray Zone," the space between war and peace. Given this, China's leaders likely did not share Crozier's view. Beijing loathes the presence of American aircraft carriers in what it views as its sphere of influence and has striven to reduce the effectiveness of American flattops like the *Theodore Roosevelt.*

Aircraft carriers are an essential component of America's power projection. There are only eleven in the fleet, spread out around the planet. Although the *TR* had docked in Guam and was offloading infected personnel, the exact condition and fighting capabilities of the American aircraft carrier was not known to America's rivals. By sending his evacuation plan outside of his chain of command, to individuals who inevitably leaked it to the press, Crozier confirmed to all of America's rivals that the US Navy operating in the Indo-Pacific was down an aircraft carrier.

At the time, the *Theodore Roosevelt* was one of only two carriers operating in the region. The other, the USS *Ronald Reagan*, had also suffered an outbreak of COVID-19 onboard. Like the *TR*, the *Reagan* was docked in port, this time in Tokyo.[9] For much of the early spring of 2020, therefore, the United States had no aircraft carriers available in the region. What's more, America's rivals – notably in Beijing – were aware of the *Reagan's* long-term presence in Tokyo Harbor. It is likely that China's military intelligence assumed the *Reagan* was combat ineffective. The *TR's* condition, however, was unknown to them. Maybe a few crewmembers were ill. Maybe the flattop was just docking at Guam as part of their mission. The cloud of uncertainty at such an uncertain time helped keep Beijing unsure as to America's strength in the region.

That is, until Crozier's letter was leaked to the press. The Crozier affair confirmed to Beijing that the Americans were debilitated, at least for a time. That period of time was a window of opportunity for Beijing's strategists to exploit. And exploit that weakness China's navy did.

In April 2020, shortly after both the *Reagan* and the *TR* were locked away in port, China's People's Liberation Army Navy (PLAN) launched a massive naval exercise. With their aircraft carrier, the *Liaoning*, sailing through the Tokyo Straits, out near the American military base at Okinawa, the Chinese flattop pivoted south where it is believed to have linked up with a group of Chinese destroyers operating in the South China Sea. Beijing timed the exercise perfectly. They recognized the Americans had been debilitated by the COVID-19 plague. They were sending a signal to the region's powers that America could not be counted on to protect them from an increasingly revanchist China.[10]

The COVID-19 pandemic worked like a charm for China. The US Navy forces assigned to patrol the Indo-Pacific and defend it from Chinese aggression were shown to be highly susceptible to a pathogen. Should an actual crisis erupt between China and the United States, and should Beijing believe that the crisis could not be resolved peacefully, China's leadership just might decide to deploy a pathogen designed to take out the crews manning America's frontline warships before China ever committed its own navy to a fight.

CHAPTER 3
BIG THINGS HAVE SMALL BEGINNINGS

Ridley Scott's ghoulish 2012 film *Prometheus* imagines a potential future wherein humanity has reached the apex of its civilization. Technological wonders, from genetic engineering and artificial intelligence to deep space travel, are common aspects of life. As the crew of the titular spacecraft lands on a strange, alien world and discovers the remnants of an advanced scientific research outpost, the protagonists come upon cannisters of alien genetic material. While playing with the strange, black, alien goop released from inside the cannisters, the film's robot (played brilliantly by Michael Fassbender) ominously mutters, "Big things have small beginnings."

What ensues is two hours of gut-wrenching storytelling which challenges the audience to think hard about what constitutes scientific progress, and if that purported progress is always good. The film's clear message is that science can be great, but if left unchecked it can threaten the natural balance of all life. Today, the world faces such a prospect. It may surprise those not aware of what's transpired in the biotechnology sector in the last fifteen years alone, but we are well and truly living in an age of rapid and robust change. And that change, while it may herald new cures for some of humanity's most troubling diseases, could also create many more complications for human life, if progress in biotechnology is left exclusively in the hands of investors and researchers.

Many of us have already lived through several radical changes in the world's scientific and technological capabilities: The advent of the nuclear bomb toward the end of the Second World War. The creation of computers and all of the subsequent applications for that incredible invention. Spaceflight. The invention of the internet. The subject of this chapter is another of these watershed changes: the 1987 discovery of repeating DNA sequences in the genomes of prokaryotes such as bacteria.[1]

A Small History of DNA

In an age when so many people believe that bigger is better and biggest is best, few realize just how supreme the small can be. As the world was in the throes of the Industrial Revolution, a Swiss scientist by the name of Friedrich Miescher first discovered DNA in the year 1869. The world looked much different in those days. The British Empire was the dominant power; steam-powered engines pushed great trains and ships around the globe; the Americans were just coming out of their horrific Civil War. But there was Miescher, ensconced in his lab, observing human white blood cells beneath his rudimentary microscope.

Miescher was not interested in the nuclein of the white blood cell ("nuclein" would eventually come to be called "nuclide" which would ultimately be known as "deoxyribonucleic acid," or simply DNA). Instead, Miescher wanted to understand better the protein components of white blood cells. In the process, Miescher accidentally observed DNA and is today rightly remembered for it. In the popular imagination, the discovery of DNA is often erroneously credited to two scientists from the 1950s: the American James Watson and the British physicist Francis Crick. In fact, what Crick and Watson discovered, after a long line of discoveries relating to DNA, was that DNA exists in a three-dimensional, double-helix formation.[2] Important, no doubt, but certainly *not* the same as discovering DNA itself.

There were several other scientists from around the world whose work led to Crick and Watson's ultimate discovery. A woman by the name of Rosalind Franklin, a contemporary of

CHAPTER 3

Crick's and Watson's, was working on innovative and important data sets that inevitably allowed for the latter pair to make their observations about the physical features of DNA. It remains controversial as to whether Crick and Watson simply forgot to credit Franklin's essential contribution to their work, or if they purposely left her out.[3]

At the moment that Crick and Watson formally discovered that DNA was a three-dimensional double-helix in 1953, the world was changed forever. Yet even before their discovery, humanity had taken interest in microbial life. Using pathogens as an invisible weapon was nothing new.[4]

In WWI, the Germans were notorious for their indiscriminate deployment of chemical weapons. Most notable was the use of mustard gas (the French used it as well), but the Germans also routinely dosed horses bound for the Allied front line with anthrax, known at the time as "glanders," hoping that the infected animals would make the enemy troops near them sick and unable to fight. Most historians have identified Germany's use of anthrax-infected horses as an early form of "biological sabotage."

In the Second World War, the Japanese possessed an advanced biological weapons (BW) program. Tokyo began in-depth research into bioweapons during the Interwar Years, giving Japan significant advantages over the other combatants in World War II (including their nominal Nazi allies).[5] Japan conducted vicious bioweapons experimentation on the conquered populations of China. The goal of such weapons is to debilitate a rival force as rapidly as possible, even if conventional, kinetic warfare is either ineffective or not an option.

Since the late nineteenth century, human beings have attempted to understand disease at the microbial level both to combat it and to weaponize it. These two aims contributed significantly to our understanding – even our discovery – of DNA. From the moment that the basic structure of DNA was discovered in 1953 by Crick and Watson, the world was revolutionized. Augmented by the concomitant advances in computing and nanotechnology (more on that later), humanity's understanding of biomedical science exploded throughout the back half of the twentieth century and into the twenty-first century, where it is now poised to become the most dynamic and prosperous industry in the world over the next decade.

But what to do with DNA?

For many years, the study of DNA remained isolated to the heady and esoteric realm of research. The real-world applications of this new science would take many years to develop. In 1987, thirty-four years after Crick and Watson determined the shape of DNA, the first human genetic map was created.[5] From that point, the quest to unlock the entirety of the human genome became the objective of biomedical researchers. According to the National Human Genome Research Institute (NHGRI), "Beginning on October 1, 1990 and completed in April 2003," the Human Genome Project (HGP) is "one of the great feats of exploration in human history."[7] The HGP was conducted by an international team of researchers whose entire goal was to sequence and map the human genome.[8] This work paved the path to today's biotech revolution.

Specifically, the exploration of DNA and the mapping of the human genome inevitably led to the study of RNA (Ribonucleic Acid). DNA and RNA are related yet different. Whereas DNA stores and transfers genetic information, RNA acts as a sort of messenger between DNA and what's known as ribosomes to help the body create proteins. A good definition of ribosomes can be found on the NHGRI website: "A ribosome is an intercellular structure made of both RNA and protein, and it is the site of protein synthesis in the cell. The ribosome reads the messenger RNA (mRNA) sequence and translates that genetic code into a specified string of amino acids, which grow into long chains that fold to form proteins." The NHGRI further defines ribosomes as being "part of the protein-generating factory in the cell. The ribosome itself is a two-subunit structure that binds to messenger RNA. And this structure acts as a docking station for the transfer RNA that contains the amino acid that will then become part of the growing polypeptide chain, which eventually becomes the protein."[9]

It is this "messenger" role for RNA that makes RNA such an essential component in biotechnology. As you will soon read, the advent of gene editing technology would not have come about had it not been for the discovery and understanding of RNA.[10]

From Small to Large: The CRISPR Revolution

The year 1987 would turn out to be a seminal year in biotech research. It was the year that the first human gene map was created. More importantly, it was the same year that Japanese scientists observed the "Clustered Regularly Interspaced Short Palindromic Repeats" (CRISPR) in a strain of *E. Coli*. From there, the rapid evolution of technology aimed at manipulating CRISPR for medical – and, potentially, military – purposes began.

Manipulating CRISPR is a relatively new concept. It could not have happened without the investment and input into overall research into DNA. It was not until 2012 that George Church, Jennifer Doudna, Emmanuelle Charpentier, and Feng Zheng discovered that they could essentially edit DNA by creating a "cut-and-paste" tool. Basically, by studying CRISPR, these scientists in 2012 figured out that by designing "Guide RNA" (gRNA) to target and isolate genes in each strand of DNA, they could effectively edit those genes to their liking. Hence, the creation of the "cut-and-paste" gene editing tool generally known as "CRISPR Associated System" (CAS).

Scientists have broken commercial CRISPR systems down into three distinct categories, types I, II, and III. Most commercial gene editing systems are type II. They include cas9, gRNA, and Homologous Recombination (HR). The COVID-19 vaccines that Pfizer and Moderna developed used gene editing. The vaccines built off "breakthroughs of the gene-editing technology known as CRISPR."[11]

The COVID-19 vaccines produced by Moderna and Pfizer are considered by most scientists to be "genetic-based" therapies, as opposed to "gene therapy," because the vaccines, when introduced to your body, do not cause your genes to change. Instead, the vaccines merely instruct your cells to produce antibodies needed to resist COVID-19.[12] It should be noted, however, that in October 2021, the president of Bayer's Pharmaceutical Division, Stefan Oelrich, proudly described his company as "taking a leap" by funding research and development of mRNA vaccines (of the kind that Moderna and Pfizer developed to combat COVID-19) that, "Ultimately . . . are an example for that cell and gene therapy." Oelrich added to the audience of six thousand assembled medical professionals from 120 countries at the World Health Summit in Berlin that year that, "I always like to say: if we had surveyed two years ago in the public – 'Would you be willing to take a gene cell therapy and inject it into your body?' – we probably would have had a 95 percent refusal rate."[13]

Oelrich is likely speaking some semblance of truth. The question is not whether the two mRNA-based COVID-19 vaccines are "genetic-based" therapy or "gene therapy." What matters is the process through which these vaccines were created. That process evolved from the observations and techniques that the likes of Doudna, Charpentier, Feng, and Church first made in their lab in 2012. And these observations and techniques will eventually be used directly to alter an individual's genes as a means of curing illness.

It is believed that vaccines and cures created through gene editing are the wave of the future.[14] These are fundamentally different from traditional vaccines. Whereas traditional vaccines put an inert germ in our bodies to trigger an immune response, the gene edited vaccines insert mRNA into the body to "teach our cells how to make a protein – or even just a piece of a protein – that triggers an immune response inside our bodies." Once the immune response occurs, antibodies are created that ultimately protect the body from the actual disease (in this case, COVID-19).[15]

CRISPR, or specifically the gene editing component of CRISPR research, has become the basis of most biotechnology commercialization today. In fact, in 2021, Big Pharma recorded a $22.7 billion surge in investment from the venture capital community into companies seeking to create regenerative therapies using CRISPR as the means of developing cures and vaccines.[16] According to Brian Gormley of the *Wall Street Journal*, biotech startups are proliferating at historic rates in just the last few years, prompting a race for patents on gene editing techniques.[17]

Therefore, biotechnology is poised to be the decisive high-tech industry of the next decade. The biotech industry and the gene editing tools they are creating are at the same point in their research and development phase that computers were at in the mid 1980s and early 1990s: the capabilities are still being tested, but biotech's potential is finally being recognized by the people with money, and key investments are being made that will ensure the biotech revolution will be as big as the computing revolution that innovators like Steve Jobs and Bill Gates pioneered.[18]

Growing Importance

On June 9, 2022, another breakthrough occurred, one that will prove to be every bit as important as the initial mapping of the human genome. MIT Professor Jonathan Weissman led a group of renowned researchers in creating "the first comprehensive map of genes that are expressed in human cells."[19] This revelation will allow for what's known as "discovery-based research" to occur. Consider this the next evolution in the quest to understand fully the human genome in order to manipulate it for medical purposes.

Thanks to Weissman and his team's research, innovation associated with the human genome can begin in earnest. We will soon be able to use this map to create vaccines and treatments according to the *specific* needs of afflicted individuals. If the Human Genome Project successfully mapped the entire human genome in 2003, then Weismann's team likely helped to create the legend for that map, telling researchers how to use it.

A Big Thing

According to author Walter Isaacson, "the biotech revolution is going to be *ten times* more important than the digital revolution, because it allows us to hack the code of life." Isaacson, who wrote a book on one of the people who discovered CRISPR, Jennifer Doudna, believes humans "shouldn't be afraid of using [CRISPR] to make ourselves healthier."[20] While Isaacson is correct in his general assumption about the opportunities inherent in gene editing and the ongoing biotech revolution, there is a disturbing trend among biotech enthusiasts and practitioners to obscure the risks posed to humanity by hacking the code of life.

For example, in her discussion with *CBS News* in 2021, Doudna expressed skepticism about concerns raised by many – such as myself – that CRISPR will be abused.[21] When confronted with the prospect that CRISPR could eventually lead to the gene editing of children, creating purportedly superior children, according to the specifications of ambitious parents, Doudna insisted that the world was many years from that becoming a possibility. Yet as we've seen with the example of He Jiankui, to which the *CBS News* report did allude, the world is clearly much closer to gene editing life for enhancement than Doudna suggests.

Any attempt to downplay these possibilities as being around the corner is irresponsible. The technology is currently rudimentary, but, given the level of interest and investment already being poured into the industry, innovation will transpire sooner than we think. With the potential profit glimmering in the eyes of innovators and investors alike, scalability becomes the key quest for all involved in biotechnology.[22] This is especially true in the wake of the COVID-19 pandemic, during which the biotech industry received a massive amount of financial investment and public interest.

Dr. Weissman's recent discovery means that the industry is much closer than anyone believed. Now that scientists will be able freely to manipulate and explore the genetic code, radical change in this dynamic industry is coming. Big things, such as the genetic engineering of the human race, have small beginnings. We are at the stage where something small is now growing into something much bigger. It might get too big to control.

Just think back to the evolution of computer technology in the twentieth century. It took about fifteen years for giant computers located in basements to be made into smaller, more complex machines. By the 1970s, the personal computer revolution happened. By the late 1980s and early 1990s, the personal computer revolution had paired with the internet, and the entire world was transformed. If, as Walter Isaacson posits in his book, biotech is *ten times* more revolutionary than the digital revolution of the late twentieth century, it is merely a question of "when" rather than "if" gene editing will be used for things such as creating designer babies and sophisticated bioweapons. Do not underestimate how rapidly these changes will come upon us.

Further, do not doubt the willingness of innovators and investors alike both to cut corners and to look for partners in the unlikeliest of places – such as the People's Republic of China. As the West increasingly looks to share information and work with China's budding biotech sector, expect the American biotech companies to take on characteristics of the Chinese Communist Party. These relationships may yield more sophisticated biotech products more quickly, but they will also represent a significant threat to the American people and our way of life. The COVID-19 pandemic may be a portent of the risks posed by untrammeled biotech development with China.

OF PANDEMICS AND CENSORSHIP

Even before the COVID-19 pandemic, the growing influence and power of Big Tech – notably the issue of censorship by social media platforms – was a hotly debated topic in the West. On the one hand, liberals believed that the Russian Federation manipulated social media platforms to help get Donald Trump elected.[1] Conservatives, on the other hand, challenged the various social media platforms for their increasingly draconian enforcement of ambiguous "community standards," arguing that these firms cracked down on right-leaning users more than left-leaning ones, thereby practicing political censorship.[2] When COVID-19 consumed the world in 2020, however, the censorship and de-platforming of people who did not conform to the narrative accepted by Big Tech (read, the Left's version of reality) went into overdrive.[3]

Importing Totalitarianism from Communist China

Egregious lockdowns meant to "slow the spread" violated many peoples' understanding of personal liberty in the United States, but they became the norm in 2020. Soon, vaccine mandates were demanded by unelected bureaucrats, imposed by ignorant and power-hungry politicians, and enforced by draconian "corporate stakeholders" in Big Tech (because all of these groups were making a profit from what Ahmed Sule called the "COVID Industrial Complex").[4] In the previously free countries of the West, the powers that be shut down speech across online platforms as part of a larger effort to curb the pandemic of fear that gripped American society. This all became part of that much-heralded "new normal," but really constituted a "diet" version of what the Communist Party of China referred to as its own "War-Time Controls," the enforced lockdown of a billion of its own people, done in the name of protecting them from COVID-19.[5]

In China, the regime sealed people in their homes and apartments, nailing the doors and boarding up windows of apartments in an infected city, all to prevent the spread of the novel coronavirus.[6] Here in the United States (and in most Western nations), while governments did not physically seal families in their homes, leaving them to die as China did repeatedly throughout the pandemic, Western governments generally tried to replicate China's approach to COVID-19 mitigation – public spaces were closed, schools were shuttered, even doctor's offices were temporarily shut down.[7] Every facet of public life was ended and placed under strict control. It was a frightening time to be an American, not only because there was a pernicious new illness afflicting our fellow citizens, but also because the government's response was onerous.

Though China's regime, which enjoys a monopoly of power in its country that few Western governments had in their own, initially managed to reduce the spread of the disease, the costs to China's economy and civil society were great.[8] Ultimately, Western nations that tried to replicate the Chinese response failed, because those Western societies are founded upon individual liberty, whereas China's is not.[9] Rather than admit defeat or even modestly adjust course to reflect the values and standards of America's democratic system, sadly, the US government has yet to acknowledge the failure of the national shelter-in-place (SIP) policies.

In China, President Xi Jinping and his government have not only refused to admit the failure of their response to COVID-19, but they've doubled down on that failure. Beginning in 2022, Shanghai, China's most prosperous and iconic modern city, was forced totally to shut down, as Wuhan was in 2020, due to the propagation of a new COVID-19 variant. The results of President Xi's "Zero COVID" policies have devastated China's economy – and might ultimately end Xi's reign, as Chinese citizens and leaders alike chafe under the draconian measures.[10]

Yet in thought, if not totally in deed, the American government's response to COVID-19 mirrored that of the totalitarian Chinese government. At the start of the pandemic in China, the regime there silenced any person who attempted to share information about what was occurring inside of China with the rest of the world – this was especially the case *before* the Chinese government officially acknowledged that they were experiencing a pandemic. There were instances of medical staff, such as frontline nurses and doctors in Wuhan, who were "disappeared" for sharing photographs or talking to Western media sources about the truth of the pandemic.[11]

China's regime infamously silenced the voice of Dr. Li Wenliang, the man widely credited as being the definitive whistleblower about the truth of COVID-19 in China. Li worked as an eye doctor in a local hospital in Wuhan. Dr. Li had first detected the novel coronavirus in his patients in late 2019. He started reaching out to colleagues of his in the area via the Chinese social media app Weibo, China's equivalent of Twitter. Using the so-called private messaging function on Weibo (really, nothing in Communist China is private), Dr. Li sought answers from his colleagues as to whether they were experiencing the phenomenon. Once he determined that the disease was pervasive throughout Wuhan, Li started trying to warn other colleagues in other parts of China, and eventually his story became international news.

Throughout his ordeal, Dr. Li's goal was not to make his government look bad – the CCP was doing that all on its own, by downplaying and denying the threat that the novel coronavirus posed its own people. Li was simply trying to work with colleagues in the field to create better responses than what was already available to him, which was not much, partly because of the information freeze that the CCP had imposed upon the Chinese medical community.

Yet once Dr. Li's messages were circulating, the local communist authorities in Wuhan came down on him. They accused him of spreading what we'd in the West call "fake news." The CCP effectively canceled him. As Li was fighting to save the lives of his patients and to keep his coworkers safe from the new plague, his government was dragging him into their police headquarters and "interrogating" him, ultimately forcing the good doctor to sign a false confession stating that he had made the whole thing up and that there was no real public health threat.

Dr. Li died of COVID-19. He was hospitalized not long after his visit by China's authorities. While in the hospital, Li refused to stop reporting on his condition. He made videos of his deteriorating condition as proof that the disease afflicting his hometown was severe – and that the world should take countermeasures against it. Li became an international celebrity. The Chinese Communist Party could not cover up his work anymore, nor could they vilify him. He had become a celebrity, largely because of his opposition to the CCP's COVID-19 policies.

So, the CCP changed course and made him into a national hero. The narrative quickly shifted away from Li as the dangerous thought criminal, and towards Li as the gallant young doctor giving up his life to save his fellow countrymen and the nation itself.[12] Yet the damage was done. Had the Chinese government worked with Li (and other Chinese medical professionals fighting to get the word out) sooner, rather than stifle the truth he was telling then, Beijing might have crafted better policies and saved more lives.

Interestingly, draconian practices similar to those the CCP imposed were forced upon the American people by their own government and by private companies (which quickly became one in the same). Whether you supported the US government's COVID prevention policies or not, US authorities shut down your places of work, schoolsing, and leisure – without much debate or concern about the well-being of the citizens who'd be disaffected by those actions. Cruises were stranded at sea for months as ports of entry were closed.[13] In some cases, such as New York City, the most severe lockdown protocols exacerbated the death toll.[14]

Meanwhile, dissenting voices were targeted by Big Tech,

and their claims shut down.[15] While the American government did not force false confessions out of or "disappear" dissenting individuals as the Chinese regime did, the ostensibly private tech firms managing social media platforms and other important aspects of the internet most certainly did digitally disappear people, via deletions of posts, de-platforming, and shadowbanning.

Just as the Communist Chinese claimed they were stopping the spread of Dr. Li's misinformation, so too were Americans digitally shut down for challenging the official line of the state. It was not only controversial public figures, such as Alex Berenson, who was censored for his contrarian reportage on the COVID-19 lockdowns and eventual vaccine.[16] Ordinary Americans posting on Facebook or Twitter were subjected to obnoxious warnings whenever they were about to read a post or make a post about information not found in the mainstream news sources. In other instances, entire social media accounts – no matter how few their followers may have been – were eradicated by capricious censors operating from the home offices of these major social media companies. Little recourse was allowed to those social media users whose accounts were terminated for purportedly spreading "misinformation" about COVID-19, the lockdowns, or the vaccines.[17]

While we should all discourage the spread of misinformation during a national emergency such as a pandemic, it is essential to remember Benjamin Franklin's admonishment that "Those who would give up essential liberty, to purchase a little temporary Safety, deserve neither liberty nor safety."[18] The Chinese system may prize repression and forced compliance as a matter of course, but in the United States, that simply should not be the case. Therefore, the willingness to stymie freedom of speech in the name of security from a pandemic was a frightening move toward the kind of totalitarianism that exists in China today.

Much of what was classified by overbearing censors in America (and throughout the West) as being "misinformation," or at the very least "misleading" information, turned out in many cases to be accurate, such as the claim that the COVID-19

CHAPTER 4

vaccines did not offer full protection against newer variants of COVID-19.[19] And, just as covering up the truth in China prolonged the suffering of the Chinese people, the silencing of truth-tellers in the United States by overzealous internet regulators hampered progress in fighting the disease from a public policy standpoint. Those who lost access to their accounts or were shadow banned were often not fully restored, even if they were proven correct over time.

Thus, the censors ultimately had the last laugh, even if they were sometimes proven incorrect on the matters having to do with the censoring in the first place. In the process, we all learned something that should have horrified us: it was very easy for our government and major tech companies to abandon any pretense of being "free" and embrace the same kind of tyranny that pervades Red China today. You will see that this dangerous pattern will be replicated again and again, and that there may be something more nefarious at play when it comes to censorship in the United States over COVID-19.

What happens when the next crisis occurs? How far will Western leaders in the public and private sectors go in responding to that crisis – how much of ourselves will we lose in that process? As you will soon read, another biological crisis is likely in the offing, given the way that Western biotech firms, scientists, investors, and even governments, through generous, taxpayer-funded research grants, willingly partner with Chinese labs and companies engaged in highly risky biotechnology research. And these Chinese biotech entities fall under the sway of China's military, the People's Liberation Army.

The questions that must be asked are, Why did the United States and so much of the West try to replicate China's response to COVID-19? Why didn't the US government and her purportedly "free" corporations refuse to follow the Chinese model? Was it sheer incompetence, or was there something more? Does it have to do with the growing relationship between Western biotech entities and Chinese-state-affiliated ones?

CHAPTER 5
THE TRUE ORIGINS OF COVID-19

To understand why Western firms and governments may have so readily embraced the Chinese model, it is important to understand that the world is still not fully aware of the true origins of the COVID-19 pathogen. While the official line has been that the infamous wet markets of Wuhan, where exotic animals are sold for human consumption, are the likely source of the disease – and thus that COVID-19, like its SARS and MERS cousins, is a naturally forming novel disease – there is much that remains unknown. Notably, Why was the Chinese government so intent on covering up the illness for as long as it did?[1]

It should be noted that China has had to contend with epidemics before. In 2003, the Severe Acute Respiratory Syndrome (SARS) engulfed China and spread globally. Just as with COVID-19 years later, the CCP's response to SARS was excoriated by the international community. But nothing was as bad as the CCP's initial response to COVID-19. In fact, after the failures of its SARS response in the early 2000s, Beijing insisted it had implemented changes to its disease prevention protocols that should have made its response to COVID-19 better than what its response to SARS had been. Of course, that too was a lie.

From the outset of the pandemic, it was rumored that COVID-19 may not have been a natural phenomenon after all. In fact, there was a contingent of people – many of them experts in the military and scientific domains – who cautioned the public that COVID-19 may have been produced in a lab. Specifically, a biosafety level 4 lab known as the Wuhan Institute of Virology (WIV), located a mere half-hour drive from the Huanan Wet Market, where China's authorities (and most of the major scientific experts in the West) claim to believe COVID-19 emerged. It seems suspect, to say the least.

In 2019, the US Department of State issued its annual report on global compliance in the Biological Weapons Convention (BWC). The BWC, first signed in 1972 by 109 different nations, today has 183 nations signed onto it, including both the United States and China. The BWC was designed to "prohibit the development, production, acquisition, transfer, stockpiling, and use of biological and toxin weapons."[2] The US Department of State does annual check-ups on the various countries (including the US) to ensure that all signatories are, in fact, complying with the terms of the BWC.

The report from 2019 is written to bore non-expert audiences. Yet the report states clearly that "Information indicates that the People's Republic of China engaged during the reporting period in biological activities with potential dual-use applications, which raises concerns regarding its compliance with the [Biological Weapons Convention]." The 2019 report concluded starkly that "The United States has compliance concerns with respect to Chinese military medical institutions' toxin research and development *because of their potential dual-use applications and their potential as a biological threat* [emphasis added]."[3] In other words, the writers of the 2019 report were saying, "We can't be sure, but it looks like China is doing some scary, possibly illegal stuff with biological research, and we think someone with greater authority in Washington should do something about it because we're just State Department bureaucrats, so we're letting you know this in a written report so that our asses are fully covered when this thing blows up in our faces – which we think it will."

Few people in Washington listened in a timely manner.

From what has been released since 2019, it sounds like elements of the Trump Administration's national security team had become interested in reports emanating from China about the true origins of the novel coronavirus. But nowhere near enough of the top leaders were clued in at a time when they might have contained the spread of the illness. Back then, if you remember, the United States was mired in an absurd impeachment scandal, America was enjoying one of the country's best economies ever, and the administration itself was more interested in ratcheting down tensions with Beijing to finalize their agricultural trade deal by year's end than it was in investigating sporadic claims of a new epidemic in China that the regime there was covering up.

There was one person in the top level of the Trump Administration who was paying attention to what was happening in China, and he spent his remaining year in government service being the Paul Revere of the pandemic: the White House deputy national security adviser, Matthew Pottinger. Mr. Pottinger was a China hawk, and his hawkishness was based on years living in and studying the country. After leaving the United States Marine Corps, Pottinger became a *Wall Street Journal* reporter covering the China beat. As deputy national security advisor, he knew the extent of lies that the Chinese regime would engage in to protect itself from the truth about the pandemic, and he intuitively understood that if the CCP was downplaying and denying the existence of the novel coronavirus, then it was going to be extremely dangerous for the world.[4] Yet getting anyone in the White House to listen initially was a problem, according to various reports on the matter.

Josh Rogin of the *Washington Post* reported that the State Department had received warnings from its sources in China about a dangerous new SARS-like illness plaguing the Chinese people in the Fall of 2019, well before COVID-19 was officially acknowledged by Beijing.[5] Even before Rogin's report was published, I had received word from a contact in the State Department's Bureau of Intelligence and Research (INR) in late October 2019 that US personnel in China were encountering what would ultimately come to be known as COVID-19. So, I knew the *Washington Post*'s reporting on this matter was accurate.

What's more, before my source in the State Department had ever said anything to me about a new illness in China, I was contacted by an old college friend, who upon graduating from DePaul University had moved to China to work in the growing investment sector there. As he told me then, he never intended on returning to live in the States again. Yet now he was contacting me, years after we had graduated, to let me know that he was moving back to the United States immediately.

When I asked him why he suddenly abandoned his freewheeling life in China, he explained that, on top of some issues that had transpired in his personal life, "there's a weird sickness floating around [in China], and I just want to get out before I can't." When I pressed him about what was going on, he said that he didn't know for sure, but that it was serious and scary – and things in China were getting bleaker as the disease spread.

CHAPTER 6
BIOLOGICAL TERROR: MADE IN CHINA

Dany Shoham is a former Israeli military intelligence officer and an expert in China's biological weapons program. He holds a doctorate in medical microbiology, and he retired from the Israeli military with the rank of lieutenant colonel in 1991. Shoham is now with the Begin-Sadat Center for Strategic Studies at the Bar Ilan University in Israel. As Shoham told my colleague Bill Gertz in the *Washington Times*, "Work on biological weapons [in China] is conducted as part of dual civilian-military research and is 'definitely covert.'" Further, it was a known fact that coronaviruses were housed at and studied by scientists at the Wuhan Institute of Virology, which is itself part of China's vast biotech research and development ecosystem, much of which falls under the control of China's National Academy of Science (CNAS), run by top leaders of the Chinese Communist Party and an integral research arm for the People's Liberation Army. Shoham did not know if the coronavirus research at the lab was directly tied to military bioweapons R&D. But since the overall Chinese bio program is dual-use (like so many other Chinese high-tech endeavors), and since the analysis of SARS being done at the Wuhan Institute of Virology did fall directly under the Chinese military's control, it is likely that there was a weaponization component to the institute's coronavirus research.[1]

Shoham was not the only expert suspicious of COVID-19's origins.

Doing Wrong in Wuhan

Even top Chinese scientists were doubtful of the wet market story that Beijing was pushing. Researchers at the South China University of Technology, a Chinese state–funded academy, wrote a widely circulated research paper positing that "[COVID-19] probably originated from a lab in Wuhan." The paper elaborates how the Wuhan Center for Disease Control (WHCDC) may have been the actual source of the pandemic. The WHCDC is located *a mere three hundred yards* from the Wuhan wet market where the CCP officially claims COVID-19 originated. The WHCDC is classified as a "Biosafety Level 2" lab, meaning its ventilation controls aren't as strict as more secure facilities (such as the Wuhan Institute of Virology).[2] The Chinese researchers assessed that COVID-19 likely originated at the WHCDC because bats were a probable transmission source for the COVID-19 pathogen into the human population of Wuhan (and, eventually, the rest of the world).[3] As it happens, the WHCDC *was* doing zoonotic research on bat-related SARS-type viruses.

In the words of Botao Xiao and Lei Xiao, the husband-and-wife team of scientists who authored the original report from the South China University of Technology, the "Genome sequences from patients [suffering from COVID-19] were 96% or 89% identical to the Bat CoV ZC45 coronavirus [originally found in the horseshoe bats the WHCDC were experimenting with]."[4] Almost as soon as the scientists had uploaded their document to the international database, it was taken down (and can now only be accessed by using internet archival searches).[5]

How much do you want to bet that the CCP's minders gave the two scientists an offer they couldn't refuse – something like: retract the report and keep a low profile until we tell you to, or disappear forever?

What's more, the WHCDC was adjacent to the Union Hospital where the first group of doctors were exposed to the novel coronavirus. Botao and Lei Xiao speculated in their now-retracted report that it was possible the disease breached its containment in the WHCDC and may have migrated over to the nearby Union Hospital, where the infection soon spread.[6] Of

BIOLOGICAL TERROR: MADE IN CHINA

35

course, this is all speculation, not because those who believe the novel coronavirus may have leaked from a lab are conspiracy theorists, but because Beijing has a long track record of covering up such horrendous events to "save face," a product of the totalitarian mindset that dominates the Chinese Communist system.

For their part, the two authors of the 2020 report have stated on their LinkedIn profiles that they retracted the original report due to a lack of specific evidence substantiating their claims.[7] Yet we know that by the time their paper was being disseminated to the international scientific community the CCP was already in the throes of a massive cover-up.[8] Remember Dr. Li Wenliang? He wasn't the only victim of the CCP's response to our generation's most threatening biological crisis to date. There were countless others, some simply reprimanded while others "disappeared" by a regime that, since long before the COVID-19 pandemic, lacked any respect for human rights.[9] After the pandemic hit, China had even less respect for those rights, especially while Chinese citizens were telling the truth behind the outbreak to the international press.

As for Botao and Lei Xiao's claims that they had to retract their papers from the international database due to the lack of available data? Well, the CCP oversaw a complete scrubbing of vital genomic databases in the weeks and days leading up to the official acknowledgment by Beijing of the presence of the COVID-19 pandemic in China. The world had been denied access to critical Chinese biomedical and genomic databases as soon as the pandemic erupted.

In fact, the WHCDC was physically relocated on December 2, 2019. In the words of the World Health Organization's (WHO) Dr. Ben Embarek, the man initially charged with determining the origin of the pandemic for WHO, early December 2019 was "the period where [COVID-19] started." Dr. Embarek is an interesting case. He was, after all, a man who spent the initial period of the pandemic shooting down any claims that COVID-19 came from a lab as nothing more than "fake news." Yet a year into the pandemic, as he led research teams attempting to determine the origins of COVID-19, he was increasingly convinced that there was something to the so-called "lab leak theory,"[10] noting that the WHCDC "moved the facility just days before the onset of the first major known case of COVID-19 on Dec. 8, 2019 – and that such relocations 'can be disruptive for the operations of any laboratory.'"[11]

Naturally, the WHCDC insists that the relocation went smoothly, and that nothing got out from its containment in the move. "Nothing to see here, folks!" But Dr. Embarek is not convinced. Beyond that, Dr. Embarek's team reported that when they had arrived in China to conduct their investigation, they were "pressured" by Chinese authorities to ignore the lab leak theory.[12]

This, of course, only led Embarek and many of his fellow WHO investigators to conclude that there must be something more to the lab leak hypothesis. Interestingly, Embarek's own agency, the WHO, attempted publicly to downplay his comments about the lab leak theory by claiming that the Danish TV documentary (in which he made his initial concerns known) had mistranslated some of Embarek's comments. Dr. Embarek, however, has been quite clear that he believes it is "highly probable" that COVID-19 infected the world because of a lab leak.[13]

The WHO has long been playing a double-game of trying to respond effectively to COVID-19 while at the same time appeasing China's leadership. The reasons for this are twofold: first, for years, China has been an integral funder of WHO operations.[14] Although some have claimed that China's funding of the WHO is not as significant as was reported in 2020, the fact is that China does have significant influence over the WHO – far more than most WHO member states. This is especially true after President Trump, in an understandable fit of rage at the WHO for its obvious downplaying of China's responsibility for the COVID-19 pandemic, pulled critical American funding for the WHO.[15] It should be noted here that Joseph R. Biden, upon becoming the forty-sixth president of the United States, restored America's support for the WHO.[16]

There's another important factor when it comes to WHO–Chinese relations. China sits atop the world's antibiotic supply chain (and other important medicines that Westerners rely upon daily).[17] Upsetting China's leadership by holding the CCP liable for the worst pandemic in decades could possibly upend the global distribution of essential antibiotics. But Embarek has yet to qualify or renounce his claims – and his suspicions are gaining traction.

In fact, the CCP–friendly leader of the WHO, the former Ethiopian minister of foreign affairs, Dr. Tedros Adhanom Ghebreyesus, is reported to have "confided" in a member of the European Parliament that he, like Dr. Embarek, believes that the pandemic came from a lab in China. Specifically, Dr. Tedros reportedly believes COVID-19 emanated from the Wuhan Institute of Virology. But like the rest of his WHO colleagues, he refuses to acknowledge this belief in public, opting instead merely to call for greater transparency from China to avoid "politicization" of the matter.[18]

Of course, if Tedros's suspicions are true (as I suspect they are), Beijing can *never* allow for greater transparency as to the origins of COVID-19. Both former Trump Administration Secretary of State Mike Pompeo as well as former Director of National Intelligence John Ratcliffe[19] are convinced that the pandemic originated in the Wuhan Institute of Virology. As you will see, China's military scientists at the Wuhan Institute of Virology were conducting intensive gain-of-function tests and other experiments on a COVID-19-like bat-born illness – with minimal concern for the safety of either their own researchers or the population outside the walls of the infamous institute.[20]

The Wuhan Institute of Virology

The mystery surrounding COVID-19's origins does not start and stop at the WHCDC. The more well-known lab leak theory centers around the mysterious Wuhan Institute of Virology (WIV). As noted previously, the Wuhan Institute of Virology was a biosafety level 4 lab (BSL-4), meaning it housed some of the deadliest pathogens known to man.

More importantly, it was involved in the *manufacture* of deadly pathogens. The WIV was the first BSL-4 of its kind on mainland China. It cost 300 million yuan ($44 million) to build, and it was constructed jointly by Chinese and French firms.

CHAPTER 6

Built far above the flood plain in Wuhan, it can withstand a magnitude-7 earthquake. The advanced lab was conceived of and approved in 2003, around the time that SARS was ravaging China, but it was not until 2014 when construction was completed. The lab itself did not begin operations until 2018. Many of the staff at the WIV were trained in Lyon, France, since Chinese scientists did not know how to manage a BSL-4 lab. The WIV's first batch of research would be conducted on SARS-like coronaviruses.

Many biosecurity experts did not believe that China needed a BSL-4-type lab. Richard Ebright, a molecular biologist at Rutgers University voicing concerns over the creation of the lab in a 2017 *Nature* article, stated that "The SARS virus has escaped from high-level containment facilities in Beijing multiple times."[21] Ebright was skeptical of the need for China to have any BSL-4 research facilities at all. This was especially the case, according to Ebright, because "Chinese researchers face less red tape than those in the West," when it comes to radical experimentation – especially with primates and other animals that usually face severe restrictions, both out of concern for an unwanted outbreak of disease and out of moral concerns for the safety of the animals being experimented on.

As Ebright lamented to *Nature* in 2017, China's quest to build multiple BSL-4 facilities, beginning with the WIV, was born more out of national pride than scientific need. What's more, a BSL-4 could be used to develop "potential bioweapons." These facilities are, by definition, dual-use labs: experiments done to create the next great vaccine to save human lives could easily be refashioned into developing the next great bioweapon. Recall the 2019 State Department report arguing that Chinese scientists were, in fact, engaged in risky biological research that violated the Biological Weapons Convention.

Another expert who spoke with *Nature* in 2017 was Tim Trevan, the founder of CHROME Biosafety and Biosecurity Consulting Services, based in Damascus, Maryland. Trevan cautioned that China's rigidly conformist and hierarchical societal structure would make operating the WIV extremely difficult. Interestingly, the Chinese National Academy of Sciences

(CNAS) representative who spoke with the author of the 2017 *Nature* article on the creation of the WIV facility asked to have their name removed from the article. It was speculated that the WIV would initiate its research into SARS-like viruses as a proof of concept; to simply test the facility's capabilities before expanding into research into more dangerous pathogens, like Ebola and the West African Lassa virus. Shockingly, the lab's exploration into SARS-like coronaviruses was considered a safe "test," since, technically speaking, conducting research into SARS-type illnesses did not require a BSL-4 facility (remember, facilities like the WHCDC were deemed BSL-2, and they were handling – or, mishandling, depending on who you ask – coronaviruses in their labs).[22]

About a year before the outbreak of COVID-19, an internal security review of the WIV by Chinese authorities found that "the lab did not meet national standards in five categories."[23] Moreover, this same panel found that "scientists were sloppy when they were handling bats."[24] We know that hundreds of bats were brought from various caves in the subtropical regions of China to the WIV for medical experimentation.[25]

Before the outbreak of COVID-19 in Wuhan, scientists conducting research on the bats at the WIV, with its apparently lax biosecurity standards (despite it being considered a BSL-4 facility), had several incidents that were highly disturbing even to those who worked there. For instance, it was reported that a bat undergoing invasive coronavirus-related tests became aggressive and attacked a Chinese researcher – which resulted in the bat's blood coming into contact with the researcher's exposed skin. That same scientist ended up quarantining himself for two weeks after another bat urinated on him, exposing him to the lab-created coronavirus they were experimenting with.[26] Another scientist, according to media reports, later complained to superiors after discovering a live tick from an infected bat outside of containment. Ticks are another excellent source of transmitting illnesses from the animal kingdom – specifically from bats – to human beings.

Since the WIV went active in 2018, China's leadership grew concerned about the ramifications of untrammeled experimentation at the lab. After all, China's scientists were untested in managing a facility like the WIV on their own. That, coupled with their rigid hierarchical culture and opaque, repressive political system, meant that engaging in risky medical experimentation could prove to be catastrophic for China. It's believed that a culture of openness with a commitment to honesty in the workplace is vital for ensuring the safe conduct of highly risky experiments. In China's system, saving face and keeping one's mouth shut is prized over all other things.

Obviously, the concerns surrounding China's creation of a BSL-4 in Wuhan were well-founded. The place was managed like a saloon in the Wild West rather than a highly sensitive research lab conducting experiments on critical viruses – any one of which, if it escaped the lab, could kill untold thousands of people and disrupt the world's economy.

According to *Nikkei Asia*, President Xi Jinping gave a speech outlining his plans for the creation of extensive biosafety laws, meant to ensure the safe operation of the BSL-4 in Wuhan. Xi's speech, given to the Chinese Communist Party's Central Planning Committee in early February of 2020, just a couple of short weeks after January 23, when Xi had ordered the initial lockdown of Wuhan and the imposition of "war-time controls" on the city to contain the outbreak, called for greater "biosafety" measures. In his speech, Xi advocated for the passage of a "biosecurity law" that would have ensured safer management of sensitive biological research facilities, such as the WIV. Xi's remarks prompted then-President Donald J. Trump to glam onto the notion that the novel coronavirus did, in fact, emerge from a Chinese lab (it was at this point that Trump began referring to COVID-19 as the "Chinese Virus").[27]

Yet many Chinese sources insist that Xi's talk was the end of a multi-year discussion within the CCP's upper echelon that began the moment that China decided to build the WIV. The discussion aimed to enact programs that would reassure the world's other advanced powers that China could run a BSL-4 of its own. Specifically, some analysts have argued that Xi's speech was in response to the case of He Jiankui, the Chinese scientist who, in 2018, gene-edited HIV cells out of two twins who were still in their mother's womb, without any approval from the state or international bioethicists. That is indeed a possibility, but it seems more like a cover for China's own disastrous management of the WIV. And, even if it were the real reason, given that COVID-19 was erupting in Wuhan when Xi gave his speech, it's clear that Xi was also speaking to dissuade the world from turning on China for its egregious conduct relating to the nascent pandemic.

The "Bat Woman" of Wuhan

The horseshoe bats that many suspect to have been the source of the outbreak were housed in the WIV for extensive coronavirus-related research programs. In fact, as early as 2005, Chinese researchers had discovered what can only be described as a precursor to COVID-19 in horseshoe bats in a cave in Hong Kong. These bats were collected for study. Similar programs had occurred throughout China since 2005.[28] At the WIV, there had been a project that created a "chimeric virus using the SARS CoV reverse genetics system."[29] Many at the time knew that if this lab-created strain leaked out into the human population, it could do immense damage.

Shi Zhengli is a noted Chinese virologist who has helped to conduct such coronavirus research at the WIV. Over the course of the pandemic, she became known in the global press as the "Bat Woman" of Wuhan. Dr. Shi earned this moniker because of "her virus-hunting expeditions in bat caves over the past sixteen years." It was her experimentation on horseshoe bats that many skeptics of the narrative that COVID-19 emerged naturally believe is the likeliest source of the pandemic. In fact, Dr. Shi even acknowledged this possibility in a rare interview she conducted with *Scientific American*, a well-respected trade publication.[30]

According to Shi, she was presenting at a medical conference in Shanghai when she received word of the outbreak of COVID-19 back in Wuhan. Shi then immediately boarded a train back to Wuhan, fearing that the disease may have come from an unsecured area of the WIV. As Shi noted to *Scientific American*,

"Her studies [of coronaviruses] had shown that the southern, subtropical provinces of Guangdong, Guangxi, and Yunnan have the greatest risk of coronaviruses jumping from animals to humans – particularly bats, a known reservoir."

But despite even her personal fears that the pandemic emerged from an improperly secured portion of her lab, Dr. Shi maintains that the genetic sequence of COVID-19 does not match any of the genetic sequences of coronaviruses she and her team were concocting. Her American colleague, ecologist Peter Daszak, the man who leads EcoHealth Alliance – a "New York City–based nonprofit research organization" dedicated to "[collaborating] with researchers, such as Shi, in thirty countries in Asia, Africa, and the Middle East to discover new viruses in wildlife" – insists that Dr. Shi "leads a world-class lab of the highest standards" at the WIV.[31]

Yet there is much controversy surrounding Dr. Shi's claims, her lab, and those in the West, such as Peter Daszak, who are rushing to her defense.

First, it is a known fact that the CCP was engaged in a comprehensive cover-up in the months leading up to Xi Jinping's public acknowledgment of the pandemic in China. In the time between when American intelligence believes the disease actually appeared in China (as early as Autumn 2019, as former Secretary of State Mike Pompeo insists) and when Xi publicly acknowledged the existence of the disease, China's authorities took drastic measures to cover up critical details.[32] Of note, the CCP destroyed genetic samples taken by Chinese scientists at the outset of the pandemic that would have allowed for the world's scientific community to determine the true origins of COVID-19.[33] Knowing the genetic code and origins of a pathogen is critical for scientists as they seek both to determine how the disease entered the human population and to craft therapeutics and vaccines to combat it.

Further, China's government is believed to have presided over the destruction of the WIV's extensive pathogen database. Alina Chan reported this disturbing fact on her popular Twitter page in 2021.[34] According to Chan, when Dr. Shi was asked by the BBC why the WIV's pathogen database was taken down, Shi

explained it "was for security reasons."[35] The French newspaper
Le Monde has proven that the database did exist but has since
been taken down.[36] A cover-up is certainly afoot. If COVID-19
was entirely natural and had nothing to do with scientific mal-
practice – or worse, military chicanery – at the WIV, why on
Earth would China seek to cover anything about the disease up,
especially when it is well known that, during the onset of a pandemic, radical transparency among the world's governments
and scientific community is essential for preventing a biological
catastrophe?

Given that the earliest genetic samples were destroyed by
Chinese scientists at the behest of their corrupt government, it
is unlikely that we can take Dr. Shi's word that the samples do
not match. If she is not lying, then it is likely that she does not
have access to the actual original genetic samples. After all, the
WIV facility is not only where advanced experimentation on
coronaviruses occurs. It is also where the Chinese military con-
ducts sensitive biological weapons experiments (under the
guise of civilian research and development programs).[37] It is
possible that Dr. Shi is being truthful when she says that she
compared COVID-19 with genetic samples of her own horseshoe
bat coronaviruses and found that no match. It's possible that
she was not part of the military operations at the WIV. But given
her stature as the leading authority on coronavirus research in
China, it is *highly unlikely* that she would not have been involved
in such an operation. Most probably she is helping to cover up
the truth of the origins of COVID-19.

If so, Shi could simply be lying for the same reason that
Botao and Lei Xiao retracted their report on the origins of the
COVID-19 pandemic in 2020: because she was threatened by her
government. There's no point in resisting a government that will
send you to a labor camp or murder you for defying it. Like these
other scientists, Shi may be under duress to spread Beijing's lie
about the origins of the pandemic.

Or she could be a willing participant in that lie, in order to
protect her own reputation and her access to the advanced
Western medical databases and expertise that she has come to
rely on throughout her career – access that would have surely

been challenged, or at least restricted, had she and her lab been found liable for the outbreak of COVID-19. If China were indeed to be found liable, the ramifications would be disastrous for the CCP and President Xi. It is unlikely that the world will ever know for certain what transpired in Wuhan – especially if Western governments and international health organizations refuse to demand total accountability from China on this pressing matter.

As for the kinds of research that occurred at the WIV, we know that the institute was not only searching for the antidotes to virulent disease. The Trump administration's Deputy National Security Adviser Matthew Pottinger reported to CBS *News* in 2021 that US intelligence had confirmed that the "Chinese military was doing classified animal experiments [at the WIV] as early as 2017."[38] That would align perfectly with the above-mentioned *Nature* article from 2017, in which biosecurity experts Dr. Richard Ebright and Tim Trevan expressed deep concerns that the WIV was going to be used to conduct risky experiments on primates and other animals that might breach containment, and that the scientists at WIV might also be conducting dangerous, "dual-use" bioweapons experiments.

Lastly, there is the added complication of Western scientists rushing to Shi's defense. Recall Peter Daszak and his Eco-Health Alliance, the non-profit in New York that is dedicated to uncovering new viruses among wildlife across the world. At first, Daszak appeared to be an authority on matters such as Dr. Shi's integrity and the judiciousness with which she and the scientists at the WIV conducted their deadly research. Alas, it has since been revealed that Daszak was hardly an impartial observer.

In a stunning revelation to the United States Congress, the National Institutes of Health sent a letter in which the organization acknowledged that it had funded a gain-of-function experiment through an intermediary – none other than Daszak's EcoHealth Alliance – that created a coronavirus that was more infectious in mice at the Wuhan Institute of Virology. In other words, it was a highly virulent, lab-created coronavirus. According to the embattled NIH, the EcoHealth Alliance had received a grant from the US taxpayer–funded government organization to conduct experiments on coronaviruses along with the WIV. Ignoring standards, the NIH claimed that EcoHealth Alliance "failed to immediately report" the results of its experiments with coronaviruses and mice. These tests involved a "limited experiment" in which mice infected with a "previously unknown bat virus [W1V1] bind to the human angiotensin-converting enzyme 2 cell receptor in a mouse model."

These mice were infected with what's known as a "chimera," specifically a coronavirus designated "SHC014" (more on that later), in which the mice being experimented on "became sicker than those infected with [the W1V1 bat coronavirus]." In essence, the NIH, through Peter Daszak's EcoHealth Alliance, was funding a freakshow in Wuhan. Many believe that COVID-19 likely derived from these NIH–funded experiments.[39]

For years, the NIH was led by Dr. Francis Collins. Collins is a man that I had respected both for his commitment to science as well as his well-known Christian faith. Sadly, in recent months, he appears to have been engaged in a high-level cover-up. Collins, along with one of his subordinates, the head of the NIH's National Institute of Allergy and Infectious Disease (NIAID), Dr. Anthony Fauci, had much to hide about the origins of COVID-19. If what the NIH begrudgingly acknowledged in its letter to Congress about its generous funding of extremely risky gain-of-function tests at the WIV is true, then America has bigger problems than merely punishing Beijing for irresponsible, possibly hostile, action with biological materials.

What this means is that Daszak's strong defense of Dr. Shi and her team at the WIV was hardly unbiased. Very likely, his support of Dr. Shi was built on strong self-interest. In that case, Daszak and the multiple Western scientists who stood alongside him in defending Dr. Shi were also assisting in the century's greatest cover-up. If they had done anything else, of course, these Western scientists would have likely been implicated in the criminal behavior that the WIV was engaged in.

Even more galling was the outsized role that Daszak played in a long-running investigation into the origins of COVID-19 organized by the *Lancet*, a preeminent science journal. That investigation was chaired by Columbia University economics

professor Jeffrey Sachs. It had to be shut down because Daszak
had been part of the investigation and, naturally, concerns arose
over his many conflicts of interest.[40] The closing down of the
Lancet investigation did grave harm to the international mission to discover the origins of COVID-19, as it effectively pulled
some of the world's foremost experts away from the investigation without yielding anything other than compromised results.

What was newsworthy about their defense of Dr. Shi and
the WIV, then, was not the fact that prominent Western scientists were vouching for China's scientific community and the
research standards exercised at the WIV. Instead, what was
newsworthy was their deep financial and research connections
to Shi and the Wuhan Institute of Virology. This isn't just Daszak
and his EcoHealth Alliance. As you will read, the bulk of the
Western scientific community has some degree of guilt for having allowed such reckless experimentation to occur in a place
like China, where safety and moral standards are so dreadfully
low. Those low standards, in fact, are what likely attracted Western investors and researchers in the first place.

During the Watergate scandal that took down President
Richard M. Nixon and his administration, his opponents were
constantly calling for investigators to "follow the money!" The
same should be asked of those investigating the origins of
COVID-19. Investigators should follow the tangled financial web
linking American and Western firms, innovators, and labs with
China, and should be prepared to discover much more than simply the origins of COVID-19.

The SHC014 coronavirus is an important link in this tale
between risky, NIH–funded research in Wuhan and dangerous
research being conducted for a period of time at the University
of North Carolina.[41] For many years, the University of North
Carolina strove to forge closer academic ties with the Wuhan
Institute of Virology.[42] This relationship persists, despite the controversy that has arisen since the pandemic erupted in Wuhan.
In 2015, researchers at UNC, along with scientists at the WIV,
"created a modified coronavirus that was shown to be able to
latch onto human cells and replicate in lung cells, efficiently
enough to cause a pandemic."[43] The SHC014 spike protein was

the key for spreading the infection in humans – and that sample was provided to Dr. Shi at the WIV by scientists at UNC.[44]

Essentially, no one should take Peter Daszak's defense of Dr. Shi and her team at WIV seriously. Daszak is merely protecting his own interests, against an increasingly angry Congress. For his part, Daszak insists that this is a giant misunderstanding; he claims, as Shi has (and the rest of the Chinese regime), that the genetic makeup of COVID-19 does not match those of the coronaviruses that his team was working with at the WIV.[45]

Yet there remains the underlying concern – how and where did COVID-19 emerge? If it was the result of risky, Western-funded research at an insecure Chinese BSL-4, then the public has a right to know. What's more, Congress must increase its oversight of the overall biotech industry. These kinds of partnerships, between Western researchers and Chinese labs, are happening all the time and could produce similar containment breaches that negatively impact the whole world.

Daszak's defense of Dr. Shi and her team was entirely self-interested. He does not want to be blamed, any more than Xi Jinping and the CCP want to be blamed – hence, the cover-up. Far from exonerating Dr. Shi and her team at the WIV, in my opinion, these facts suggest that Peter Daszak and much of the West's scientific community are accomplices to the crime that likely occurred in Wuhan.

CHAPTER 7
GAIN-OF-FUNCTION TESTS

The WIV was engaged in a process known as a "gain-of-function" test. These tests are deemed critical by researchers, as they allow for scientists to study the origins, evolution, and modes of infection that potential illnesses – notably, pernicious coronaviruses – take, contributing to the medical community's quest to create vaccines for such illnesses.[1] But these gain-of-function tests are risky under the best circumstances. In the Chinese bio lab, where rapid results were favored above all other considerations (like safety or morality), the risk was far greater than what it should have been.

If, as I suspect, there is truth to the lab leak theory, then China is liable for damages to the world. But China may not be exclusively liable. After all, American scientists – many of them working for or with the US taxpayer–funded National Institutes of Health (NIH) – helped to teach China's scientists in Wuhan how to conduct these risky gain-of-function tests.[2] And, as you will read in later chapters, American and Western scientists and investors are helping China's overall biotech sector utilize CRISPR-Cas-9 gene editing technologies to develop advanced biotech *that could eventually be used in a conflict against the United States.*

The NIH had been conducting gain-of-function tests for years. The objective was to allow scientists to learn the genetic code and the pattern of development for illnesses to preemptively create vaccines. Coronaviruses, such as COVID-19, SARS, and Middle East Respiratory Sickness (MERS), have plagued humanity.[3] For years, fears have abounded that the next great pandemic would likely come from this family of viruses. Understandably, scientists wanted to map out the most likely next pandemic *before* the pandemic came about, to create a vaccine in a rapid timeframe (thereby reducing mortality rates). Yet critics of this process have long maintained that its risks far outweigh any possible benefits.[4]

In 2012, for example, the National Science Advisory Board for Biosecurity (NSABB) recommended to the US government that a "voluntary pause of sixty days on any research involving highly pathogenic avian influenza H5N1 viruses leading to the generation of viruses that are more transmissible in animals."[5] The temporary moratorium was enacted after the NSABB reviewed two manuscripts written by scientists who had conducted risky gain-of-function experimentation on the H5N1 virus in their labs. In those manuscripts were "molecular information on mutations which allowed H5N1 to become transmissible in mammals."[6] Something about the H5N1 gain-of-function experiments in these manuscripts spooked the members of the NSABB enough that they not only requested a pause on all gain-of-function research nationwide, but also redacted the information found in these manuscripts.

Once the sixty-day ban on gain-of-function tests lapsed in 2012, the Obama Administration stepped in and "proposed an indefinite continuation of the moratorium on gain-of-function studies with H5N1 viruses that could affect mammalian virulence and transmissibility" until the scientific community and government could agree on what constituted acceptable and unacceptable risk.[7] They did agree on something: Ralph Baric's controversial gain-of-function tests on the SHC014 coronavirus found in horseshoe bats should go forward. Despite the moratorium having been imposed by the US government, Baric's work on the SHC014 coronavirus gene began before that moratorium was enacted by the Obama Administration. Baric's study was essentially grandfathered in so that the money that was spent on the risky study would not have been for nothing.

Plus, the NIH jiggered with the definitions of a "gain-of-function" so that Baric's study could be exempted without prejudice from regulators. As you've seen throughout this work, both Chinese and American scientists have little qualms about playing fast and loose with rules meant to protect vulnerable populations from the ravages of science experiments gone awry, and with key definitions of scientific activity. Frankly, there was simply too much money involved in discovering new vaccines for the scientific community to close up shop.

Dr. Francis Collins, when asked about why the NIH had imposed the moratorium on gain-of-function tests in 2013, had this to say about the matter: "These studies, however, also entail biosafety and biosecurity risks, which need to be understood better." He said this as the NIH allowed for Baric's studies to move forward. Collins's NIH then altered the definition of "gain-of-function test" to ensure that the SHC014 study went ahead, despite the moratorium that former President Obama's government had imposed.

Even after the SHC014 experiments in the United States ended, Peter Daszak's EcoHealth Alliance stepped in to continue the research at the Wuhan Institute of Virology under the leadership of Dr. Shi Zhengli. Dr. Shi had the horseshoe bats while both Baric and Daszak had figured out how to collect SHC014 from them. Together, with Fauci and Collins's funding, they all experimented on SHC014 to make it transmissible to humans.

Interestingly, the paper where Peter Daszak and Shi Zhengli detailed their findings from the 2018 tests done at the WIV acknowledge that the NIAID "jointly funded" their research along with the National Natural Science Foundation of China. Also supporting the researchers was a PREDICT Project Grant from the Emerging Pandemic Threats program of the United States Agency for International Development (USAID).[8] What were these US government organizations thinking by funding biotech experiments at a Chinese lab that was suspected of conducting "dual-use" bioweapons experiments? And what, by the way, did the scientific community gain from these very dangerous experiments? According to Shi Zhengli, Peter Daszak, et al., their gain-of-function experiments yielded the conclusion that,

"Our surveillance [of potentially harmful diseases in the animal kingdom] is not exhaustive.... more extensive surveillance in this region is warranted."[9] That conclusion hardly seems worth the headache.

For the record, Peter Daszak defended Baric's gain-of-function study at the University of North Carolina back in 2015, as his organization, EcoHealth Alliance, was raking in funding from the NIH to collect samples from animals for further gain-of-function tests.[10] As Savio Rodrigues detailed in his exclusive reporting at the *Sunday Guardian*, an Indian newspaper, Dr. Shi, Peter Daszak, and Ralph Baric exchanged a bevy of emails at the start of the outbreak of the novel coronavirus in Wuhan in which the scientists fretted over the designation of the virus.

Specifically, Dr. Shi worried that virologists calling the disease the "Wuhan Pneumonia" (or some variation thereof) as a shorthand reference would "stigmatize" the region from where it emanated. There was further fear that naming it the "Wuhan Pneumonia" would encourage critics to blame China for the outbreak. The two American scientists wholeheartedly agreed with Dr. Shi and endeavored, through their contacts with the NIH and the international scientific community, to ensure that no official designation for COVID-19 gave any credence to the regional origins of the illness.[11] Obviously, if Wuhan were to be blamed, at some point, Baric and Daszak would also possibly be blamed – as would the whole enterprise of working on risky biotech experimentation in China.

As others have noted publicly, however, not referring to the illness in public discussions by its regional origin was a break from previous practices. To name just one example, few could tell you about A(H1N1) and its historical importance.[12] But almost everyone has heard of the disease through its popular name, the "Spanish Flu."[13]

The email exchanges that Rodrigues revealed in his reporting clearly indicate a concerted effort by Shi, Baric, and Daszak to get the name of the illness changed so as not to encourage ordinary people to connect Wuhan and the novel coronavirus. This was part of a larger self-defense public relations campaign waged by the three scientists. It was all about managing perceptions and shaping public opinion to keep the world's prying eyes away from their NIH–funded, Chinese Communist Party–backed, risky research shenanigans in Wuhan. Why would these scientists have otherwise cared if it was referred to as the "Wuhan Pneumonia"?

According to Shi Zhengli's emails with Baric and Daszak, her own team in China had taken to calling it the "Wuhan Pneumonia" because it was easier to remember than the coded string of letters and numbers. Names and labels are useful because they are meant to cut through and clarify something complex. Yet when one purposely avoids these conventions, when one instead compels the use of a complex name that creates uncertainty and confusion, then one is not interested in truth. One is merely attempting to obfuscate.

The cover-up was not merely relegated to China, where deceit is an essential component of the government. The outright lies and absurd obfuscations were as infectious as the pandemic itself. It propagated from the CCP to the international scientific community, and from the scientific community to national and global organizations such as the National Institutes of Health and the World Health Organization. The cover-up spread beyond these areas, too, to wider society, as noted earlier in this work.

While speaking to a military audience in Florida, as the pandemic was spreading globally (just before the national lockdowns in the United States commenced), I began giving my breakdown on Western investment into China's biotech sector. I'd given the lecture dozens of times over the previous two years to military audiences. With what was then the still-barely understood COVID-19 phenomenon, I decided to incorporate what I knew about the illness into the lecture. Using academic research papers referencing gain-of-function tests and, specifically, the works of Ralph Baric, I briefed the military audience on what I knew.

I then veered into reporting from sources who were working in the field to determine the genome sequence of COVID. By that early point in the pandemic, reliable information was scarce, but there was a smattering of reports that indicated certain existing medicines used to fight HIV and Ebola, for example, were helping patients afflicted with the novel coronavirus.[14] I then cited the fact that the COVID-19 genome had mutations related to HIV and Ebola, that my contacts in the scientific community were referring to the disease as a "chimera."[15] I explained to my audience of US military intelligence professionals that, as far as the scientists working on the origins of the novel coronavirus could tell, COVID-19 was unlike most other coronaviruses. It reminded one of the SHC014 pathogen – which, as you remember, was described by scientists as a "chimera" – that Ralph Baric, Peter Daszac, and Shi Zhengli had been working on for several years before the pandemic occurred.

And as soon as I said those words, the individual in charge of the unit I was speaking with grew apoplectic. After the event, as I was waiting at the airport to return home, I was contacted by the gentleman who organizes these talks for me. It turns out, the colonel was irate with me. This colonel insisted that I write a formal apology and explanation for my statements. I stand by everything I said. The colonel then banned me from the base, though I have since been asked to return. Officially, the colonel claimed I had made a homophobic comment because I had said that HIV medications were helping to ameliorate the symptoms of some patients afflicted with COVID-19. How that was a "homophobic comment" is beyond me.

More likely is that I was talking about something that only people in the intelligence community were privy to at that time. I had to be silenced for speaking out of turn. And I was, for almost a year. While I would be asked to give more lectures at other facilities, with different audiences, on a myriad of geotechnology subjects, it would be more than two years before I would ever be asked to report on my research in the biotech field. By that point, everything that I had said to that initial military audience in 2020 was already out in the public – and being attacked and critiqued by people like Peter Daszak and their supporters, in both America and China.

Of course, everything I said was factually true. At the very least, there were scientists around the world who were saying the same things I was saying. Point in fact, not more than a few months after I had made those comments to the military audience, multiple sources had come out with analyses of the COVID-19 genomic sequence.

Yes, there were HIV genes present.[16] As well as Ebola, cancer genes, and even Malaria (which is why some antimalarial medicines were helpful in dealing with the symptoms of people infected with COVID-19).[17] The presence of HIV genes indicated to many scientists that there was foul play involved, because these were not normally found in naturally occurring coronaviruses, such as SARS or MERS.

The prevailing wisdom among the global scientific community was that the presence of these genes, while unique, did not confirm the illness definitely came from a lab. Yet their presence did not preclude the possibility either. Considering that China spent an inordinate amount of time and resources covering up the outbreak of the disease and removing key genomic databases from the internet, and that the SHC014 pathogen created via gain-of-function tests was a chimera of similar makeup to COVID-19, it seems a plausible scenario indeed.

Luc Montagnier, the legendary French virologist who was awarded the Nobel Prize for his co-discovery of HIV, and who has since passed away of old age, insisted that the presence of HIV genes within the COVID-19 genome sequence proves the disease was lab created. He repeatedly referred in public to the pandemic as resulting from an "industrial accident," his term for an accidental lab leak.

Few other scientists agreed with Montagnier.[18] Montagnier had become controversial in his old age because he was a strident opponent of vaccines and he had become convinced that DNA emitted electromagnetic energy.[19] Without getting into the finer details, his work was deemed "quackery" because he based his claims on a purportedly disproven theory of "water memory."[20] Montagnier believed that by using electromagnetic waves to "blast" away the HIV genes in COVID-19, they could render the illness harmless to humans, largely because the HIV and Malaria elements were artificially fused to the coronavirus genome to create this chimeric illness. Regardless of whether one agrees with the body of Montagnier's work, or his proposed solution to the COVID-19 pandemic, there remains much controversy around the fact that HIV and Malaria were present in the genome of COVID-19.

In November 2021, after lecturing a group of military leaders in California on China's rapid development of advanced technology, I was pulled aside by one of the audience members. I had just finished speaking about China's technological development, discussing biotechnology as one key area where China was moving ahead of the United States, and I had made a brief reference to my theories on COVID-19. For the first time, no one pushed back on my claims. I was curious about this experience.

It became clear why no one in the audience challenged my assertion. This man told me that most of his colleagues believed COVID came from a lab. The gentleman worked in US military intelligence. He informed me that he specialized in analyzing high-technology development.

On a slightly related note, an academic paper was published that indicates a strong correlation between COVID-19 outbreaks and the presence of fifth-generation (5G) internet infrastructure and usage in both China and the United States – indicating that there is at least a possibility that there exists some interplay between illness, genetics, and electromagnetic waves.[21] Lastly, several scientists have postulated attacking the COVID-19 disease electromagnetically, by using electromagnetic waves to disrupt the illness's ability to bind to human cells or to disrupt its natural processes on the EM spectrum.[22][23] So, while these solutions may be radical, they are not far off from what Montagnier was talking about when he was urging his fellow scientists to look at electromagnetic waves as a possible solution to the illness (although the scientists pushing for these solutions do not appear to endorse Montagnier's more controversial theory of "water memory").

But such inconvenient facts have never stopped those seeking to protect the status quo from punishing those who think differently. Anyone daring to ask these questions is immediately digitally canceled by Big Tech in the United States or, in China, physically disappeared or intimidated by government bureaucrats who have something to hide.

CHAPTER 8
WHO BENEFITED FROM COVID-19?

Since the outbreak of the pandemic in 2020, there have been more than six million deaths (and counting) worldwide.[1] What followed as a result of government policies in response to the pandemic caused the worst global economic downturn since the Great Depression.[2] In September 2022, President Joe Biden proclaimed that the pandemic "was over."[3] In typical fashion, President Biden's claims were promptly walked back by a senior administration official – specifically by Dr. Anthony Fauci – because the disease continued to kill people.[4] We still don't know the origins of the disease, and at this rate we are unlikely ever to do so, simply because too many powerful people both in China and in the West stand to lose money, prestige, and power from the public learning the truth. Yet, what *is* known is that pharmaceutical companies made windfall profits.[5] In fact, as a result of the emergency economic policies implemented by the government to combat the negative impact that COVID-19 had on America's socioeconomic system, the wealthy and powerful got wealthier and more powerful.[6]

People must now, three years after the worst pandemic in our recent history, ask themselves two related questions: who benefited from COVID-19, and how did they benefit?

It was not just Big Pharma that benefited from the pandemic. Big Tech companies benefited, too. Companies such as Google, specifically its parent company, Alphabet,[7] made bank.

Amazon, with its cloud computing department which partnered with vaccine-maker Moderna to enhance its digital information sharing; enriched itself during the pandemic.[8] Microsoft, which invested considerable sums to partner with vaccine producers to better the COVID-19 vaccine rollout, made loads of money as a result of the disease.[9] And don't forget that Facebook (yes, Mark Zuckerberg funded GlaxoSmithKline's search for a COVID cure in 2020 to the tune of $100 million) enjoyed great financial gains. Because these Big Tech firms massively diversified their investments, reaching deep into biotech development, they all got wealthier and more powerful during a biological crisis. It's no longer only about faster computers or better apps for Big Tech. These Big Tech companies, with their massive profits, are investing in multiple, variegated industries.[10] Conglomerates like Alphabet are less concerned with their core technological services, and function more akin to a bank.[11]

This, of course, is not automatically bad. Yet it does create inherent conflicts of interest, between those companies charged with managing public platforms, like a major search engine or a large social media company, and their newfound investments in COVID-19 countermeasures. These Big Tech companies had a vested interest in "trusting the science," and thus in stifling free discourse, throughout the pandemic and beyond. Your protestations online, when taken together with the skepticism of major public figures or even companies challenging the vaccine requirements, represented a direct threat to the bottom lines of Big Tech in America. Your questioning of the origins of COVID-19 was a direct threat to those researchers who had everything to lose if gain-of-function experimentation was completely terminated, not just here in the West, but in China as well.

And once former President Donald Trump and his followers started questioning the origins of COVID-19, with so many of his followers questioning the efficacy of the lockdowns, social distancing measures, and even the vaccines themselves, all bets were off. Big Tech, which was already predisposed to disproportionately punish conservative voices online (especially those which supported Trump), happily involved itself in the ongoing debate. Because Big Tech owned the social media platforms and search engines that everyone uses, these companies enjoyed outsized power to act as regulators of the digital public square. Big Tech used its power to demonize and destroy any person or group who dared to challenge the accepted wisdom. Big Tech came down on the side of Big Government's scientists, Big Pharma, and even the Communist Party of China against ordinary Americans.

It began with that one small kernel of inquiry: "where did COVID-19 originate?" Within two years' time, because so few in the scientific community were being honest in their answers to this question, the conversation had evolved to encompass deeper – and still unanswered – questions: Who in government can we trust? To what degree is the integrity of America's biotech sector compromised by China? What will Big Government and Big Tech do to protect their own malfeasance and corrupt interests? Can you really believe that the vaccines are good for you and your children?

Lies regarding the origins of COVID-19 were sold by the likes of the CCP and its partners in the Western scientific community, who clearly had something to hide. Those lies were bought hook, line, and sinker by the dupes (and, in some cases, willing accomplices) in Western media, government, and business. Today, the institutions that were designed to better protect the American people from disease, whether the NIH or WHO, to say nothing of the companies that produce medicines we all depend upon, have lost significant amounts of the public's trust. Their apparent willingness to silence critics as they force us to accept controversial policies damages that trust still further. This does not bode well for any of us, as these institutions are essential. And these essential institutions have lost that trust because they stopped respecting individual rights and liberty, the cornerstones of American civil society, and because they have started aping (badly, I might add) the draconian practices of totalitarian China.

Rather than standing up as a beacon of truth and probity in a time of severe biological crisis, of resisting the siren song of authoritarian deception, many (if not all) of the West's most prominent scientists, biotech investors, and research facilities accepted the distortions of the CCP. Across the board, the US government, its scientific community, and its private corporations engaged in what Eamonn Fingleton, author of the incredible 2008 book *In the Jaws of the Dragon: America's Fate in the Coming Era of Chinese Hegemony*, would describe as "reverse convergence," whereby the free institutions of the West began emulating Communist China rather than encouraging Communist China to emulate them.[12]

Consider this: even former President Barack Obama's administration believed that gain-of-function tests were both too risky and unnecessary for the furtherance of medical science. The NIH – which reports to the president – opted essentially to ignore the wishes of its political leadership and to continue supporting risky experiments even after it had been explicitly ordered to stop. Finally, when things got too dicey for the researchers to continue their experiments in the United States, they offloaded the program to China, where standards were far lower. And the Chinese were happy to conduct these experiments under the imprimatur of scientific research, as China gained sophisticated biotech R&D which the Chinese military could then fold into its increasingly sophisticated bioweapons program (which ultimately threatens its primary geopolitical rival, the United States).

Scientists in favor of gain-of-function tests would have you believe that these are effective ways for crafting vaccines and fighting diseases which plague mankind. But many prominent scientists oppose gain-of-function tests (such as those who cautioned the US National Science Advisory Board for Biosecurity against them in 2012) because of the great risks associated with them. Simon Wain-Hobson of the Pasteur Institute of Paris is a leading opponent of such research. He has been vilified since initially opposing the gain-of-function tests that Baric, et al., were conducting in 2015. Since that time, his opposition has only been proven prescient.[13]

Following that unfortunate series of events, a transnational cover-up began, involving the Chinese Communist Party, the international scientific community, Big Pharma, and Big Tech – along with their willing accomplices in the "mainstream" media.

CHAPTER 8

The aim of the conspiracy was both to protect the dangerous gain-of-function research and to ensure that massive profits were enjoyed by Big Pharma and Big Tech. Our biosecurity and freedoms – to say nothing of the truth – were sacrificed in the process. If we aren't careful, soon those same gain-of-function tests could be brought back to American shores, as most scientists believe they are necessary (when they clearly are not). Even if they aren't brought back to the United States, there's nothing stopping the biotech industry from continuing them in another country, such as China, because so many powerful actors in the scientific community are convinced that the route to vaccines lies in such dangerous experiments.

Now we must contend with an even starker possibility: that this lab-created disease did not accidentally leak out at all. What if, God forbid, it was a bioweapon that was launched intentionally? And even if it wasn't, as so many experts insist, what's to stop China from weaponizing its extensive biotechnology industry and using such weapons for political and military gain?

Coronaviruses can be "artificially manipulated into an emerging human disease virus, then weaponized and unleashed in a way never seen before," wrote eighteen PLA officers in a shocking 2015 document entitled *The Unnatural Origins of SARS and New Species of Man-Made Viruses as Genetic Bioweapons*.[1] The next time someone tries to condescend you about how and why it is scientifically impossible not only for COVID-19 to have come from a lab, but for it to have been a bioweapon, just repeat those damning words from China's top military scientists. If that doesn't change your interlocutor's mind on the matter, then that person is likely a bad faith actor in this ongoing discussion. It should be noted that, after having initially discounted the "lab leak theory," the US intelligence community reopened its investigation into the origins of COVID-19 in 2021.[2]

China's "Goddess of War" Takes Charge of COVID-19 Response

Chen Wei is a major general in the People's Liberation Army of China, its leading expert in biowarfare. She commands a great deal of respect, not only in China, but throughout the international scientific community. Major General Chen helped to develop a nasal spray to assist with the symptoms of SARS during the outbreak in China in 2003.[3] During the 2014 Ebola outbreak, Chen gained notoriety again for her contributions to the global efforts to combat that outbreak.

Her stature has grown in recent years as she has taken the commanding heights of sophisticated biotechnology research and development for the PLA. After the COVID-19 outbreak in Wuhan and Beijing's implementation of "war-time controls" to slow the spread of the pandemic, Major General Chen was placed in command of the effort to contain the spread and develop a vaccine. For her efforts she was awarded the national honorary title of "People's Hero."[4] In some military circles, however, she's known by a much more ominous moniker: "The Goddess of War."[5]

There remains much controversy surrounding her control of China's military forces during the outbreak of COVID-19 in Wuhan. Officially, she and the military only took control of the situation once it had reached crisis levels. Until early 2020, we are told by China's government, the military took a backseat to the civilians running Wuhan. But this is hardly something that is agreed upon, given that the regime there imposes harsh information controls to prevent Western media sources from discovering a litany of truths about the situation in China.

The Wuhan Institute of Virology, which had effectively shut down its operations once people began asking the tough questions about the origins of COVID-19, was reactivated under General Chen's command. It was here that the PLA's military scientists were supposedly working along with the "Bat Woman" of Wuhan, Dr. Shi Zhengli, to develop China's indigenously produced COVID-19 vaccine. Critics of Beijing's official narrative about the origins of COVID-19 have been quick to point out that China's scientists were disturbingly slow to pinpoint the purportedly natural origins of the illness – in stark contrast to the typically short four months it took for Chinese scientists to identify the origins of SARS, for example. Yet the development of a COVID-19 vaccine in China for public purposes has been done at record speed.

From the moment General Chen assumed command of the WIV on January 25, 2020, her work began in earnest on a vaccine. Using the immense resources at her disposal, Chen's team had a single-shot COVID-19 vaccine, produced by a Chinese state-owned firm known as CanSinoBio Biologics, Inc., at Phase I trials in mid-March 2020 – a mere two months after General Chen landed in Wuhan. They moved from Phase I human trials to Phase II a few weeks thereafter.[6] It must be stressed that, if the lab leak hypothesis about the origins of COVID-19 is indeed true, this record timeframe for getting a usable vaccine makes sense. They already had access to the true genetic data of the disease, which, according to this hypothesis, was created in their lab, and stored in that genetic database that Dr. Shi had slyly taken off-line before COVID-19 was officially announced as a pandemic to the world.

We know that Dr. Shi moved rapidly in the early weeks and months of the pandemic to take the WIV's extensive genomic database on the gain-of-function tests they had been running offline.[7] Labs that were not directly affiliated with the PLA in China were ordered to destroy their genetic samples of COVID-19 as well. Anyone who could not be brought onboard with the narrative that Beijing was crafting to deflect blame away from China's freewheeling biotech sector was silenced or, worse, disappeared.[8] Wuhan itself was not only locked down but scrubbed of any evidence that international investigators might have been able to find about the true origins of the disease.[9] Then, the PLA took direct control of the area. With their access to the WIV's extensive (now closed) database, General Chen and Dr. Shi were likely able to figure out which genetic template COVID-19 was based on and how best to formulate a vaccine. In this context, it makes sense that China would create a vaccine so quickly.

According to most medical experts, the Chinese vaccines are nowhere near as effective as the Western versions.[10] But the Chinese versions shouldn't be written off entirely.[11] China's scientists are actively perfecting them. In fact, General Chen is now enhancing her original single-shot vaccine by creating a COVID-19 booster that is inhalable. Whether it is as effective as the Western versions of the vaccines currently on the market remains to be seen. But the speed of the creation of the Chinese vaccines is suspect. As for why their vaccines wouldn't be as good as the Western versions: remember, China only just got the BSL-4 capability a few years ago – and their first major operation, testing coronaviruses, ended in a global pandemic.

General Chen was an expert in creating the very kind of chimeric diseases in her labs that I (and many others) believe the COVID-19 pathogen represents. As my colleague Steven W. Mosher has argued, the COVID-19 genomic sequence is unlike that of any naturally occurring disease. COVID-19 is so infectious to humans because it "burrows its way into human cells using a special tool called a 'Furin cleavage site,'" Mosher explained to his readers in the *New York Post*: "A new scientific report shows that of the 1,000 – one thousand! – coronaviruses found in nature that most closely resemble the novel coronavirus that caused COVID-19, not a single one possesses a similar 'furin cleavage site.'"[12]

According to the scientific paper that Mosher referenced in his 2021 article, not only was COVID-19 many times more infectious than other coronaviruses found in nature, but the presence of this furin cleavage site has led scientists to seek unconventional treatments for the disease. For example, the authors of the paper tested "4,000 compounds including approved drugs and natural products" to see what could counteract the presence of this furin cleavage site in the COVID-19 genome. Extraordinarily, they found that "the anti-parasitic drug, diminazene, showed the highest inhibition effects on furin." The authors recommended that the anti-parasitic drug be used as a countermeasure against COVID-19.[13]

To be clear: this is the sort of tinkering around with natural pathogens to make them more virulent to humans that would only be found in laboratories – specifically, labs conducting biological experiments in that uncomfortable gray area between gain-of-function research and bioweapons development. Hence, why General Chen is a key figure in this tragic saga.

The prevailing wisdom, even among skeptics of China's official origin story for COVID-19, is that the coronavirus accidentally leaked from the WIV facility. Once it slipped out from the facility, this theory goes, the Chinese Communist regime went into overdrive to cover up their mistake. One cannot help, however, but to wonder if there wasn't something even more nefarious going on. While evidence supporting the lab leak hypothesis is scarce, it does exist. Determining if China purposely launched COVID-19 as some type of biological attack will be nearly impossible. But, if you'll allow me, I'd like to spend a few pages conjecturing.

Bio Agent of Chaos: The Fear Pandemic

Standing before an enthusiastic crowd of supporters in Indiana, during the controversial 2016 presidential campaign, Donald Trump raged against the People's Republic of China. The soon-to-be forty-fifth president argued that China had been "engaged in the greatest theft in the history of the world" against the United States.[14] Trump accused China of manipulating its currency to make its exports more competitive globally. He then added this zinger to the event: "We can't continue to allow China to rape our country!"[15] The Manhattan real estate mogul–turned–reality television star then assured his enthralled audience that, if elected, he'd "turn it around" with China. After all, in Trump's estimation, "we have the cards … we have a lot of power with China." Toward that end, the 2016 Trump campaign promise was to "cut a better deal with China that helps American businesses and workers compete."[16]

One of the first controversies of the Trump Administration – even before Trump had been sworn in as president, during the transition period from the Obama presidency to Trump's – was his first call with a foreign leader. At the urging of former Republican Senate Majority Leader (and one-time Republican Party presidential candidate) Bob Dole, Trump spoke with the pro-independence leader of Taiwan, President Tsai-Ing Wen.[17] This move by Trump infuriated Beijing. A few weeks after the Trump call with Taiwan's leader on November 8, 2016, the Chinese Navy captured a US Navy drone operating in international waters.[18] Basically, Beijing was doing its best to test the new, incoming American president, especially after his hostile remarks about China throughout his campaign, and his breaking of diplomatic norms by calling the Taiwanese president.

Once Trump became president in January 2017, he went to work fulfilling his ambitious trade policy against China. Trump was fully committed to getting American and Western companies to move out of China and return to the United States, in a process called "on-shoring." Trump also started a trade war with China in which he began slapping onerous tariffs on all agricultural goods coming from the United States and going to China. This, more than anything else Trump said or did against China, was a major red flag for the Communist leadership in Beijing.

Writing in 2009, preeminent geopolitical theorist George Friedman cautioned that China's greatest fear was an American president who would implement a program of trade protectionism.[19] China, after all, depended upon imports from abroad. While China has become an advanced country with the world's second-largest economy, it still has problems with acquiring the resources it needs to survive.[20] Food, energy, essential ores – the very building blocks of a modern economy – these things have been historically scarce in China. And this fact has proven to be a strategic vulnerability for Beijing that US war planners, like former Marine Corps Colonel T. X. Hammes[21] as well as Mackubin Thomas Owens, have advocated exploiting.[22]

Food scarcity has often led to regime changes in China's long history.[23] When former President Trump began tariffing US agricultural exports – notably the soybean – going to China, in an effort to force China's leadership to come to the negotiating table, it is likely that many Chinese leaders, including President Xi Jinping, saw this as more than just jockeying for deals.[24] They likely saw it as a declaration of war through other means.[25] This was precisely what the Chinese had spent the last fifty years doing to America and the West: waging an onerous war through other means.[26]

Some analysts in the West believed that the Trump Administration's trade war was a total disaster.[27] It did have some unintended negative consequences for the United States. Notably, the trade war harmed American farmers, who depended on

China as a large (and growing) export market for their goods.[28] It hurt American manufacturers as well, since US agriculture companies sold tons of tractors and other farm equipment to China. By placing tariffs on those goods, President Trump caused prices to skyrocket and slowed their trade with China, hurting both farmers and firms alike. Incidentally, many of these people and industries were essential Trump supporters going into the 2020 presidential campaign.

But despite these negative effects, Beijing ultimately did come to the table.[29] To compound matters for President Xi Jinping, Hong Kong was erupting while Trump was squeezing China with his trade war. From Beijing, the view was bleak: food was becoming more expensive, being threatened by a reinvigorated, more hostile America as led by Trump, despite the denials by Western media sources clearly intent on declaring Trump's trade war a loss before it was even finished.[30]

Then, the denizens of the dynamic city of Hong Kong commenced protesting against Beijing's new "National Security" Law that most Hong Kongers rightly believed was an attempt to stifle their democracy. Again, Xi Jinping and his ministers likely viewed this as part of a far-reaching conspiracy by the Americans to destroy the CCP without waging a war.

Beijing entered negotiations with Trump from a position of weakness – something that China's rulers have not done in decades when it came to dealing with Americans on trade. They'd effectively gutted the American Midwest of essential manufacturing jobs over the last forty years and were in the process of pilfering high-tech jobs and other white-collar industries from the West via one-sided trade deals. They did so, that is, until Trump came to power.

China has mastered the art of what's known as civil–military fusion.[31] This is exactly like it sounds: the marriage of the private sector and government, specifically the military, into one, cohesive unit – with the CCP sitting atop the whole thing. It is a permanent wartime economy primed at defeating all threats, foreign and domestic, and dedicated to pushing aside any challenger on the international stage, readying China's eventual rise as the world's greatest superpower. American business, political, and scientific leaders believed that after China opened itself to trade with the United States in the 1970s, the two powers became allies. This was not how the leadership of the CCP saw things. China's leaders looked upon the United States as food to be devoured to make China the dominant world power by 2049, the centennial anniversary of the founding of the People's Republic of China by Mao Zedong.[32]

Trade policy with America, therefore, was not viewed by Beijing as merely the negotiated settlement of mutually beneficial economic policies. Instead, Beijing looked at trade with the United States (and the West) as extractive: they needed advanced manufacturing capabilities to build a robust middle-class in China that would, in turn, empower China's rise into a modern, high-tech superpower that could challenge – and defeat – the United States. Beijing's communist rulers became adept at using American corporate leaders' greed, as well as the greed and naivete of American politicians, to their advantage.

In exchange for immense short-term profit, American leaders gave the store to China. Harvard Business School and most other elite institutions in the West looked favorably upon trade with China as a "win-win" that would ultimately convert China into a vibrant democracy permanently aligned with the United States and its Western values. But Chinese rulers saw the ostensibly "free" trade as a chance to supplant America as the world's leader.

This explains why Beijing's leaders freaked out when Trump initiated his trade war. Trump wanted a better deal, and he targeted China's critical importation of foodstuffs from the United States to do it. In short order unrest fomented in Hong Kong, necessitating a military crackdown, and pressure on China on the world stage mounted.

Xi Jinping did not see these as unconnected events. As the ultimate embodiment of will-to-power in China, President Xi saw the evil hand of an orange American tyrant who was more attuned to Beijing's nefarious behavior than any US leader in decades. To Xi, Trump had declared war on China by going after its essential food supply. Xi thought Trump meant silently to force regime change in Beijing, because this was exactly the kind of thing Xi would do as leader of China. Xi sensed vulnerability everywhere. He needed to strike back at his obnoxious American rival, Donald Trump.

But how?

China's military, despite undergoing historic and rapid modernization, was not ready for a traditional war with America. What's more, China still needed to do business with the West, so attacking the Americans or their allies directly was a bad idea. Besides, Xi Jinping knew that most members of the American (and Western) elite, regardless of political party, loathed Donald Trump. It is possible that Xi and his strategists determined the best way to "nudge" America into a position that would benefit Beijing *without* risking a wider kinetic war with the US military would be by launching a biological attack on the US.

Dr. Sharad S. Chauhan of the *Indian Defence Review* described China's sophisticated Wuhan Institute of Virology as the "aircraft carrier of virology," and he reminded his readers that the French intelligence services opposed the decision of Paris to partner with Beijing in building China's BSL-4 in Wuhan, when China's leaders first proposed it to France during the SARS outbreak of 2003–04. According to French intelligence at the time, there was no doubt that China would repurpose their BSL-4 to conduct biological weapons experiments. France's Directorate General for External Security [DGSE] was concerned that it could not confirm that the technology the Lyon-based firm, RTV, was giving China for the construction of the WIV would be used solely for civilian research purposes. According to Chauhan, "[RTV] depended on an organization, the China National Equipment of Machinery Corporation (CNEMC), *which was controlled by the People's Liberation Army* [emphasis added]" to verify that French technology was not being abused by China for covert bioweapons research.[33]

According to Dr. Chauhan, the WIV took fifteen years to construct simply because China consistently broke its promises to France about not weaponizing the facility. At every turn, the French would have to come in and verify that China was not planning to convert what was supposed to be a civilian BSL-4 into a bioweapons facility – and the French had difficulty confirming anything. French intelligence later reported that the Chinese were pulling back from their original agreement entirely, after construction on the WIV had already been nearly completed. Is there any wonder why?

Yet the French continued building the facility (likely viewing it as a sunk cost and because Beijing probably greased the wheels, the way it always does when dealing with Western governments). Given the statements of senior Chinese leaders over the last decade, Beijing clearly believes that weaponized biotechnology is the silver bullet its forces need to wage war indirectly against America and its allies.

You've already read how PLA Major General Chen Wei, the so-called "Goddess of War," was placed in command of the WIV during the pandemic. Some have rumored that she was pulling the strings well before that. Why place a military general in charge of the response to a disease? Certainly, in times of crisis here in the United States, military officers have been given positions of leadership Rarely, however, are they made the ultimate authority in a time of domestic crisis. Why General Chen? While she was a leading authority in biological research, her real area of expertise was *weaponizing* pathogens. If this was meant to be a civilian response to a civilian crisis, why not appoint someone like "the Bat Woman of Wuhan," Dr. Shi Zhengli?

This was a military operation through and through. US Army Major Joseph Murphy, while serving at the Defense Advanced Research Projects Agency (DARPA), "objectively analysed [sic] activities planned (and performed) by the Wuhan Institute of Virology in conjunction with various collateral information [and concluded] that the initial virus generating the pandemic was created at WIV, and accidentally leaked therefrom in 2019."[34] Dr. Lawrence Sellin, a retired US Army colonel, noted that, "in 2019, [SARS-CoV-2] was sent from the PLA Eastern Theater Command in Nanjing, its headquarters, to WIV, specifically for testing on monkeys."[35] The WIV and various PLA personnel and programs worked closely together to "enhance bat coronaviruses adaptation to humans (thus leading to SARS-CoV-2)," according to Israeli biodefense expert Dany Shoham.[36]

Regardless of those rumors, though, Chen Wei's standing in

the military, and the speed by which she achieved that standing (as well as her previous medical innovations), indicate that she plays an outsized role in China's overall biotech sector. This would accord with the extensive research demonstrating deep connections between China's military and civilian programs. The table, therefore, was set for the whole COVID-19 affair to have been a covert PLA operation from start to finish.

This cuts to the core of what civil–military fusion is. In the West, notably the United States, we have a strong view of keeping separate the public and private sectors – if not always in practice, certainly in theory. Even when Big Government and Big Business cooperate, they do so usually in limited, specific areas, and the results tend not to be as efficient as in more independent private sector ventures. In China, whatever lip-service Beijing may pay to creating a strong private sector, they remain communists at heart and recognize that the only way to compete against the Americans is to fuse their civilian and military sectors. This civil–military fusion has yielded incredible results for Beijing. In just fifty years, under this model, China went from being an agrarian backwater to the second-largest economy in the world in GDP terms and the largest in PPP terms. Draw back the curtain, however, and it's the PLA that is pulling the strings. And the PLA reports to the technocrats who command the Chinese Communist Party.

Major General Chen was heavily involved in the research of bat coronaviruses.[37] As noted previously, she was instrumental for China during both the SARS epidemic and the Ebola outbreak in 2014–15. It was her unit, along with Dr. Shi Zhengli at the WIV, that experimented on bat coronaviruses and endeavored to make them highly virulent to humans via gain-of-function tests that greedy, ambitious, and ignorant American scientists helped to fund and organize via generous NIH grants. General Chen was again called upon to "manage" the outbreak in Wuhan and to develop the Chinese vaccines for the disease, which she did.

But the question remains: to what strategic end does making COVID-19 as a bioweapon – and deploying it against both the world and your own population – serve?

CHAPTER 9

The New Total War

Think back to what was going on from 2017–19. China was losing face on the world stage. Trump was doing the one thing Chinese leaders had assumed no American president would do, and the one thing they feared: initiate a trade war. That Trump targeted China's food supply, famine being the historical impetus for regime change throughout Chinese history, was an added insult and concern. This, coupled with the unrest in Hong Kong, must have sent shivers down Xi Jinping's spine. While it's true that the American "victory" in the trade war with China was not that great, Trump was showing his people – and future American presidents – what can be accomplished with enough gumption and a willingness to play all of America's cards in trade. Xi had to reset things in his favor.

After all, the Americans did beat China in the agricultural trade war, in that Chinese companies that endured Trump's tariffs sold far less to the United States than they would have, had Trump's trade policies not been in effect. What's more, China's GDP loss as a result of the trade war was "three times as high as the U.S.," according to estimates from Yang Zhou, an economist at China's prestigious Fudan University.[38] Despite what critics of Trump argued, China did ultimately cave to Trump. China's leaders met former President Trump and signed on for a new trade deal, one that was meant to reduce China's advantages in the agricultural sector.[39] No matter how small of a victory it was for the United States, it was a victory, nonetheless. Imagine if the Americans started doing this on every major issue with Beijing. Xi Jinping's back would be broken over time.

But Xi and his ministers understood that Trump's grip on power was tenuous at best. Not only was he surrounded by enemies, but Trump was also an unexploded bombshell in the field – he was just as likely to harm his own political fortunes as those of his enemies. The point of Xi Jinping unleashing a bio-attack on the West would have undoubtedly been to remove the real threat in Washington: Donald Trump. That the United States was moving into a presidential election year, when

COVID-19 conveniently erupted in China, would not have been missed by Xi in this scenario.

Yet seen one way, in December 2019, Trump appeared unstoppable. The US economy was the strongest it had been in decades. The president was pressing his "Make America Great Again" agenda on all fronts with powerful effect, despite the opposition of his political rivals, who looked smaller every day. He had survived impeachment. This larger-than-life figure loomed over Beijing and Xi Jinping.

Removing Trump reset the great game in China's favor. The American people would do it themselves. Allowing for the leak of a virulent pathogen that was deadly enough to induce planetary panic but not enough to destroy China's economy and political system was a perfect way to achieve this goal. Yes, it was risky. But so long as Trump and his administration were running Washington, China's presence on the world stage was severely threatened.

The novel coronavirus played perfectly into the communists' hands. First, it was a proof-of-concept for China's PLA: if the plan worked, it would be a new means of attack to use on the West. The Western powers would have little defense against this, the ultimate silent killer, a new stealth agent in the global shadow war. Next, it allowed for the CCP to increase control at home and use "war-time controls" to silence any perceived critics of the regime. It allowed for Beijing to put its populace in a total war mentality. The COVID-19 pathogen disproportionately killed the weak and the elderly – two groups that China's rulers historically view as burdens on their society and its readiness for war. Then, the pandemic, once it reached the West, eviscerated the free and open societies there. The pandemic not only destabilized the political order in the United States – which would have been the goal of the operation, if COVID-19 was, in fact, a biological weapons attack – it also killed the American economy.

As the United States and the rest of the West struggled under the weight of the pandemic, China was able to weather the worst aspects of the disease. My colleague at the *Asia Times*,

David P. Goldman, outlined in vivid detail how China's rulers used COVID-19 as an opportunity to both expand their Orwellian surveillance state as well as to test new advanced technologies to assist in disease mitigation. Artificial Intelligence, using 5G wireless networks and Huawei smartphones that most Chinese citizens owned, was deployed with great effect to identify and isolate individuals who might be infected.[40] At the expansive Chinese manufacturing plants on which the rest of the world is far too dependent for critical goods, sophisticated automated robots were employed to augment the reduced capabilities of human workers during the height of the pandemic.[41]

Indeed, the pandemic (at least in the beginning) worked in the favor of the CCP's economy. China re-opened its economy before the Americans could open theirs.[42] In summer of 2020, as the United States was in the doldrums of its pandemic response, China's GDP growth was stellar, much better than that of the United States.[43] A year later, its economy posted positive gains. Many believe as a result that China could possess the world's largest economy in nominal GDP terms as early as 2028.[44] Beijing appeared to live up to the fact that, in Chinese, the word for "crisis" is composed of two characters, one representing "danger" and the other "opportunity." Plus, if the solution to the pandemic were to increase central control, it makes sense that a highly centralized society, with vicious autocrats running the country through an Orwellian security system, would do better than the relatively individualistic, open, and democratic societies of the West.[45] It was a made-to-order crisis for the communists in China.

By 2022, Hong Kong had been mostly crushed beneath the jackboots of the CCP.[46] Xi's political enemies were being targeted, even as Xi himself, though still firmly in power, was facing stiffer political opposition at the top levels of the CCP for going possibly too far with his COVID-19 response. The United States had been shaken to its very core by the pandemic: there was regular talk of a pending civil war in the United States between the Left and Right; the economy was the worst it had been in decades; and a sclerotic, potentially dementia-addled septuagenarian, Joe

Biden, who was likely in hock to the CCP on some level, had replaced the dynamic and tough Donald Trump.[47]

With the rise of the pandemic went any semblance of the world order that existed before it. The year 2020 was essentially Year Zero in international affairs. America had been hobbled, and China had been empowered. So, too, had other American rivals, such as Russia, who opted to take advantage of the chaos to push into Ukraine in February 2022.

The Russo–Ukrainian War was hugely destabilizing for the West. It forced the Americans to rededicate their limited global military power away from Asia toward Europe (giving Beijing a window of opportunity), while the Russian invasion into Ukraine, a major agricultural power, caused the price of food-stuffs to rise around the world. At the same time, America imposed a harsh sanctions regime upon Russia for its illegal invasion of Ukraine, which took a large portion of fossil fuel sources off the global market, causing the price of energy to rise for everyone – further contributing to America's economic slump.

And, as that happened, Russia became closer to China as never before – so much so that what had once been idle talk of a new world order led by China and Russia was fast becoming a bleak reality. Much of the energy that Russia would have sold to the West, after America got its allies to follow along with its onerous sanctions regime on Russian energy, ended up being sold to China.[48] Plus, Russia's excess agricultural capacity (it is an agricultural superpower) was sold over to China, which sat just across a large and porous land border with Russia.[49] With America's forces distracted by the sideshow in Europe and the Biden Administration crying "havoc!" and letting slip the dogs of war in Europe, ostensibly to protect Ukraine, the Indo-Pacific was left wide open to Chinese aggression – just as it had been during the twenty years of wasteful and inconclusive wars in the Middle East the Americans had gotten bogged down with after 9/11.

This all occurred because of a microscopic pathogen that was likely produced by Chinese military scientists in Wuhan, with the financial and intellectual support of American scientists and biotech innovators. Even today, it is a source of controversy to point out that China likely created COVID-19 in a lab. For those who are daring enough to say this, still fewer of them are willing to entertain notions that this may have been a deliberate act of aggression by China's leadership against America's. But it's hard to overlook the implications of China's COVID-19 bio attack, if it was, in fact, a biological weapons attack.

Xi Jinping had a bee in his bonnet from 2017–19: Donald Trump. So long as the Orange Man was in the White House, the United States was a significant, direct, and growing threat to China. Once Trump was removed from power, though, the Americans went back to the way they were from the 1970s to 2016. Historically, American leaders were ambivalent about, even sometimes encouraging of, China's rise. Overnight, the US went from Trump's promise to "Make America Great Again" back to the Barack Obama mentality of "managing America's decline," with Joe Biden's election to the presidency.[50]

What's scarier was how quickly Trump's fellow Americans piled on. Essentially, China created a knife in COVID-19, which Xi Jinping happily placed in the hands of Trump's vicious political enemies at home. Before the pandemic hit the United States, there was nothing that could stop Trump and no one who could replace him as president in 2020. Once the novel coronavirus washed across America's bucolic shores, however, all bets were off. The pandemic turned everything upside down.

With the economy collapsing and fear over the invisible enemy reigning supreme, Trump had nothing to offer a scared American people, who were being further gaslit into hating the Republicans by a biased and vengeful mainstream media machine that worked in conjunction with the Democratic Party. The media not only strove to eradicate any public support for Trump, but also wanted to replace that public support for Trump with support for a bumbling, aging buffoon, Joe Biden, who under normal conditions, such as the previous two times he ran, could have never been elected as president.

Further, as elements of America's own scientific community and government were intimately involved in supporting the WIV – specifically, bat coronavirus gain-of-function tests –

suddenly our public health policy response became weaponized
politically. The cover-up gripping China soon came to grip the
United States. Too many powerful people had too much to lose
if the truth got out about what precisely had been going on
inside the WIV.

America's elites recognized COVID-19 as an opportunity to
finally rid themselves of the meddlesome forty-fifth president.
It wasn't just the elected leaders of America who were inimical
to Trump. It was also the permanent bureaucracy – "The Swamp,"
as Trump had termed it during the 2016 campaign. Though they
worked for the president, most of the bureaucrats in Washington, D.C., hated President Trump. This was not only true of those
in the intelligence community, a constant bugaboo of the forty-
fifth president. It was also true of the health bureaucracies
charged with protecting America from a pandemic.

During the pandemic, there was concern that the Trump
White House would meddle in the Food and Drug Administration's (FDA) vaccine approval process. Addressing those concerns
in a joking manner, the then–Assistant Secretary of Health and
Human Services for Public Affairs, Michael Caputo, and the
Trump Administration's Commissioner of Food and Drugs,
Stephen Hahn, observed, "Government scientists would sooner
climb the roof at the FDA's White Oak offices in suburban Maryland and light themselves on fire than take political action."
Caputo reportedly added, "None of those electric cars in the
parking lot at White Oak have Trump bumper stickers."[51] Hahn
and Caputo were really saying that the medical bureaucracy in
Washington didn't want to do anything that might enhance the
Republican Party's standing in 2020. They were all Democrats
opposed to Trump, the most controversial Republican leader
in decades.

Drs. Anthony Fauci and Francis Collins of the NIH were the
two primary leaders appointed to run the US response to
COVID-19. As you've read, these two men had funded risky gain-
of-function research, both in the United States and, later, at the
WIV in China. These were the people who supposedly served
President Trump. They were neither friends nor allies, judging
from their private remarks and actions.

In this scenario, former President Trump is much like Caesar (if I may evoke the words of Oliver Stone as they related to JFK): he is surrounded by enemies, and something is underway, but it has no face. In this scenario, Fauci, Collins, et al. are *not* coordinating with Xi Jinping. They don't have to. Their interests simply aligned (as did the interests of the media, Big Tech, and the Democratic Party). These stakeholders are responding to the initial input from China – COVID-19 – and moving quickly and decisively to defend their interests and further their agenda – all of which happens to gel nicely with Beijing's own anti-Trump agenda.

The pandemic was the most serious public health crisis in many of our lifetimes. But when compared to other pandemics in the past, it does not hold a candle.[52] As my friend and colleague Gregory Copley wrote in 2020, COVID-19 was primarily a fear pandemic.[53] The fear of the disease triggered a cascade of responses from the world's authorities that, in the case of the United States, was entirely out of character. What's more, the fear pandemic was so serious that it triggered a series of knock-on responses that reversed the positive trajectory that the United States had been riding, putting the country instead on a path of terminal decline. And, more frighteningly, potentially toward a catastrophic world war – one that it might lose.

Thus, if COVID-19 was a bioweapons attack, it was the perfect weapon. Yes, it killed many people, but it was nothing like previous pandemics. The attack had significant political impact that redounded to the benefit of the CCP. It was, in effect, China's "Biological 9/11" perpetrated upon the United States. But instead of triggering a military response from America, as Bin Laden's 9/11 did, the US response to COVID-19 aimed inwards.

Lastly, consider the response of some of America's closest allies to China in the aftermath of COVID-19. Almost immediately after the pandemic struck the world, Australia, a nation that had concerned Washington because of its growing ties with China, completely reversed course. In Canberra, Australian leaders who were once totally enthralled by the promise of wealth via trade with China began scrambling for ways to put distance between the two countries' economies. Many Australian leaders began publicly denouncing China for their role in the pandemic. In fact, most Australian leaders believed that the COVID-19 pathogen did come from a lab and began publicly demanding that Beijing allow for an international investigation into the WIV.[54] This has obviously caused a great deal of tension in a relationship that Beijing once valued highly.[55]

About whether COVID-19 was a bioweapon, Dr. Mark Kortepeter, a bioweapons expert with years of experience working with the United States Army, concluded in 2020 that, regardless of COVID-19's true origin, the pandemic has "reminded us of our vulnerabilities as a society" and that he had little doubt that "our adversaries have been taking notes on how challenging it has been for the US to respond effectively" to the pandemic. Kortepeter ominously concludes, "It is only a matter of time until we face this type of challenge again – either from mother nature or an adversary. Now is the time to shore up the vulnerabilities in our preparedness and response that this pandemic has laid bare."[56]

If COVID-19 was a bioweapon attack, it was probably a proof-of-concept; a test our enemies were using to prepare for the time when they were ready to launch something truly catastrophic; a weapon against which our society had little defense and that might give China the advantages it needed to defeat the United States and reorder the world system to their liking. Dr. Kortepeter, therefore, is correct: the bioweapons implications of the pandemic are too strong to ignore, and time is not on America's side to prepare itself. Judging from how poorly the United States responded to COVID-19, to how readily America's elites adopted totalitarian strategies, the future looks bleak indeed should we be subject to another pandemic or an actual bioweapons attack from China.

As was noted earlier, the same labs involved with risky gain-of-function research on coronavirus vaccines in China for the last several years are now spearheading extensive research into developing a smallpox vaccine. If COVID-19 emanated from a Chinese lab that had been developing a vaccine for coronaviruses, then people should be fearful of what China is doing with smallpox. Whereas COVID-19 was a nuisance to the global system, smallpox is one of the deadliest pathogens known to man. Few people today are inoculated against it. What's more, smallpox has long been a fixation of bioweapons developers because of its virulence and lethality.

Thus, if it hasn't come already in the form of the COVID-19 pandemic, society must be prepared for a Chinese-produced Biological 9/11, possibly involving smallpox.

Canada's National Microbiology Lab (NML) is in the quiet town of Winnipeg, capital of the bucolic Manitoba province. The lab has the prestige of being Canada's only BSL-4 laboratory. Although it is Canada's sole BSL-4, the facility has a long and storied history of disease research. And, since it is in Canada, the NML enjoys a long partnership with the United States' National Institutes of Health, its Centers for Disease Control, and several other important American medical research arms, both in the public and private sectors. The BSL-4 in Winnipeg is a major source of international biotech research and development, meaning the facility not only impacts Canada and the rest of North America, but is an essential cog in the global biotechnology research and development machine. The NML employs scientists from around the world, including the People's Republic of China.

Winnipeg's National Microbiology Lab is well respected because of the important role the facility and its employees played in the global response to the Ebola outbreak in Africa in 2014. If you'll remember, the world's powers – as led by President Barack Obama – responded by immediately deploying copious resources to the outbreak sites in Africa, generating much controversy at the time. When I was working in government, the Obama-led response was criticized heavily for, as many leaders believed, needlessly putting American (and Canadian) scientists' lives at risk by potentially exposing them to the illness, and for elevating the risk that these individuals might accidentally bring the disease back to North America. People were concerned, too, that the US government was wasting money it didn't have and limited amounts of equipment it'd possibly need for when an illness erupted in the United States. Yet the Obama Administration was correct: by addressing the illness at the source, they likely curbed the threat of a pandemic.

Canada assisted in the global effort. In fact, it was Canada's NML that pioneered the use of "monoclonal" antibodies, which scientists believed could be used to prevent Ebola (and eventually other illnesses, as we've now seen in their use against COVID-19) from infecting healthy cells in people. The scientists of the NML's Special Pathogens Program had been working on ways to treat a multiplicity of related illnesses, from HIV to coronaviruses to Ebola. One of the most promising members of the Special Pathogens Program was a young doctor and biologist from China, Xiangguo Qiu, who joined the team along with her husband, an HIV expert and fellow biologist, Keding Cheng, in 2003. Monoclonal antibodies, at that time, had only been used to treat cases of cancer, but Xiangguo believed they could be scaled up to treat a coterie of viral infections.

The challenge for Xiangguo and her teammates was that Ebola tended to swamp the immune systems of a patient. That meant the window of time during which medical intervention must take place was compressed. Dr. Gary Kobinger was the man in charge of the Special Pathogens Program at the NML. Few believed that monoclonal antibodies could be used to treat such a rapid, vicious viral infection before it killed the patient. According to Dr. Kobinger, "Despite the fact that everybody was saying that it [would] never work, she kept going."[1] Her perseverance ultimately paid off: during the Ebola outbreak of 2014–16, the NML spearheaded the creation of effective monoclonal antibodies. Xiangguo's solution was to overwhelm the body of a person infected with Ebola by introducing three different monoclonal antibodies at the same time. The treatment protocol became known as "ZMAb" (it was named by Xiangguo).

ZMAb proved instrumental in stemming the Ebola outbreak. For her hard work and dedication, Xiangguo was awarded a Governor General's Innovation Award in 2018. Upon receiving the award alongside Dr. Kobinger, Xiangguo Qiu proclaimed that, "It's not just that we found a cure for Ebola, but our work is having an impact on the whole scientific community. It has become a blueprint for treating those other infectious diseases." With multiple publications and a major award under her belt, Xiangguo had the world in front of her. Yet few people outside the scientific community had ever heard of her.

Sadly, if she is remembered now by a wider public, it is for the caustic events that transpired in the NML in 2019. At that time, members of the Royal Canadian Mounted Police (RCMP) stormed into Xiangguo Qiu's lab and shuttered it. The RCMP then confiscated the scientist's computer and any data she had been working on. Both she and her husband were detained by the RCMP. They were ultimately fired in 2021 from their roles under a cloud of suspicion and controversy. All their research and work was stopped indefinitely. While the RCMP has not explained why it seized Xiangguo Qiu and her husband and impounded their work, most analysts believe that they were evicted from the NML for having violated Canadian national security law. Despite these strange circumstances, many of their colleagues at the NML insist that the fate of the scientific power couple from China was a "bureaucratic snafu."[2]

Given the positive contributions made by Xiangguo Qiu and her husband, it seems unlikely that the Canadians would arbitrarily remove them from their positions and kick them out of their scientific community. While government bureaucracy can be inefficient, mercurial, and plain wrong on any given day, the removal of the pair did not just happen randomly. The two were flagged by Canadian security and their removal was the end of a long-running investigation by RCMP and other elements of Canada's security services into suspicious connections between the pair and their home country of China.[3] Because the Canadian government has not been forthcoming with the details of what caused the removal of the two from the NML, it is always possible that Xiangguo Qiu and her husband are victims. Yet there is plenty of circumstantial evidence that points at least to extreme negligence on their part. At most, it points to an active, long-running conspiracy on their part to offload sensitive biological material and research from the NML to the WIV.

Here is what's known. During their highly acclaimed work for Canada's NML, both Xiangguo Qiu and her husband sought increasing levels of cooperation between their lab and the Wuhan Institute of Virology.[4] In fact, many of their colleagues at the NML encouraged these links, believing them to be an essential sharing of the burden of expensive, time-consuming research. The partnership, many believed, would cut down on the time it would take to develop critical vaccines and treatments for pernicious ailments. Yet some people – notably members of the Conservative Party in Canada – were concerned by the growing connections between the NML and the WIV. Like the French, who ultimately helped China to build the WIV, Canadian security services and the Conservative Party believed that Canada sharing sensitive lab samples, research data, and techniques would not only give China the opportunity to pilfer Canadian intellectual property, but also facilitate China's use of the data for the creation of bioweapons.

Xiangguo Qiu and her husband were the connective tissue linking Canada's NML and China's WIV. Without them, the partnership and growing proximity between the two facilities would likely not have occurred when and how it did. In 2018, the relationship cooled, and Ottawa began questioning its policy of close ties with China. It is likely that the reassessment by the Canadian government of its overall friendship with China brought Xiangguo Qiu and her husband under closer scrutiny. And whatever Canadian investigators found troubled them enough to remove them from their positions at the NML.

According to public documents, the couple "shared information and samples of Ebola and Nipah [viruses] with the Wuhan Institute of Virology after the Winnipeg lab agreed to send the samples to the Chinese lab" in May 2019.[5] For the record, the US government believes that Ebola and Nipah are bioterrorism agents and highly restricts as well as closely monitors their use in scientific experimentation.[6] There is a great and reasonable desire to keep these pathogens from proliferating to bad actors, because they can so easily be weaponized. In fact, Xiangguo's colleague at the NML (who defended her and her husband) told *MacLean's* that Xiangguo "was sometimes guilty of playing fast-and-loose with the rules" governing the sharing of information.[7]

That's bad enough, if she was just arrogantly concerned with cutting corners to achieve maximum results on risky bio-tech research and development. There's plenty of that mentality in the scientific community to go around. But the pair's professional connections to China are extensive. Specifically, their relationship with none other than PLA Major General Chen Wei lends itself to the notion that their removal from the NML was not merely a "bureaucratic snafu," as many of their former colleagues insist.[8]

In May of 2018, one year before her removal, Xiangguo Qiu wrote an email to her supervisor asking if she needed to sign an "MTA" (Material Transfer Agreement) to send samples of Ebola from the NML to the WIV. In her email, Xiangguo made an understandable lament: she dreaded having to go through the endless paperwork just to send a measly sample of Ebola for what many believed was mutually beneficial research, simply because that research would be done in China. Her supervisor commiserated with her complaints about the bureaucracy, but replied that she nevertheless had to file an MTA and get approval for the transfer. Even though the MTAs were drawn up and Xiangguo received approval for shipping the samples to the WIV, the MTAs, according to *MacLean's*, went unsigned. And that lack of signature from an authority granting ultimate approval for the shipment is what likely triggered the initial investigation that eventuated in the termination of Xiangguo and her husband.

If that were the only incident that precipitated the firing of the two scientists, one might be inclined to feel sympathy for them as victims of a faceless, thoughtless Canadian bureau-cracy. But why were the research assistants from China that the two had brought in as lab assistants also removed? The mostly Chinese students the pair had hired as staff had come under suspicion of espionage, too.[9]

CHAPTER 10

And what of Xiangguo's ties to Major General Chen Wei? The two collaborated on multiple scientific research papers. The defenders of Xiangguo and Keding Cheng downplay these connections. But the papers in question were on the pathogens that ultimately contributed to what many believe was the creation of COVID-19 at the Wuhan Institute of Virology. The relationship was years old and would have only come about because Xiangguo and her husband were sharing proprietary, dangerous samples and research from the NML with the PLA–managed WIV. Xiangguo Qiu and Major General Chen Wei (who wrote many of these academic papers under a pseudonym), spent much time conducting research on Ebola. Data was shared between the NML and the WIV – often with the encouragement from scientists at the NML, who wanted to conduct their research more quickly.[10]

One cannot be an agent of espionage without first making sure one's bills are in order. According to scientists at the NML, Xiangguo and Keding Cheng had purchased their "dream home" shortly before they were fired from the NML.[11] The house was appraised at $1.2 million. The couple owned a secondary property which they rented out. The two had another property, a cottage worth $524,000, in a small town in Manitoba called Gimli. Xiangguo and Keding are also believed to own a mansion in China.[12] While they may have legitimately purchased the mansion in China, the CCP routinely rewards their "national heroes" with material wealth and benefits for having done a great national service – like propelling China's biotech and bioweapons programs by granting them access to advanced Western research.

Since being axed by the NML, no one has been able to locate the controversial couple. Rumors abound that they are back in China, living in their mansion.[13] It has been confirmed by Canadian journalists that the couple has neither sold their "dream home" residence in Canada, nor are they living there any longer. For such an expensive home, how can it sit empty for so long? Reporters from Canada's *Globe and Mail* went to their house in the early summer of 2022 and determined that their two adult sons had been living in the property.[14] Whether the children are paying rent to cover the costs of the mortgage for their parents is unknown.

Where are a couple of government scientists getting the money to cover such costs? While it's true that doctors and medical researchers make good money, those who work for government labs, such as the NML, usually make less than their private sector counterparts. Together, Xiangguo and her husband, Keding, made low-six-figures. How were they able to afford all that property, with the requisite down payments? Often, foreign agents are supported significantly by their governments when they are on mission. This is done to allow for the agent to focus exclusively on his work. It is likely that, on top of whatever monies Xiangguo and Keding earned for their legitimate work through the NML, they were also being supported via the Chinese government for their support of China's overall biotech development.

I do not believe Xiangguo Qiu and her husband created COVID-19. Rather, I believe they offloaded essential samples to the WIV and worked closely with scientists there to study Ebola. There was another program running within the WIV directly tied to Major General Chen Wei and the Chinese military's biological weapons program, and Xiangguo Qiu's information on Ebola was folded into this secret program. Chinese high-tech espionage never happens all at once, from a single source. It rarely appears to threaten the target country's national security directly. It is done over a protracted period and in piecemeal fashion, to avoid suspicion.[15]

No single researcher has the total image of what's going on. It is unlikely that Xiangguo Qiu and her husband directly helped the gain-of-function research going on with coronaviruses at the WIV. It is, however, important to note that the COVID-19 genomic sequence shares similarities to the genes that comprise the Ebola virus.[16] So, yes, it is probable that Xiangguo Qiu's research indirectly contributed to gain-of-function experiments at the WIV. And it shouldn't be overlooked that Xiangguo's husband, Keding, was a leading HIV researcher (as you've read, some believe that there are unique inserts in the COVID-19 DNA that share traits with HIV-1). It remains to be determined if their assistance was done out of a naïve commitment to scientific discovery, or if Xiangguo Qiu and her husband were active and knowing participants in a high-end espionage operation.

But the bizarre financial circumstances of the married couple, in addition to the fact that they maintained a secret residence in China, that Xiangguo Qiu wrote countless academic papers with Major General Chen Wei, and that Xiangguo and Keding brought scores of Chinese students to the University of Manitoba and then into the NML facility as research assistants for years, are suspect, at the very least. Accessing the NML facility requires top-secret clearance. The Chinese student research assistants would not have been given those clearances without the support of Xiangguo Qiu and Keding Cheng.

It turned out that many had concealed ties to either the People's Liberation Army or Chinese labs. One of the Chinese graduate students that Xiangguo brought to the NML was Feihu Yan, of the People's Liberation Army's Academy of Military Medical Sciences. The Public Health Agency of Canada (PHAC), which oversees the NML, denies that Yan ever worked at the NML. Yet researchers at the NML confirmed that Yan was a fixture at the lab when Xiangguo worked there and that both Yan and Xiangguo coauthored eight scientific papers together.[17] These facts lead one to the stark conclusion that an espionage operation probably was underway and the husband–wife team at the NML were knee-deep in it – no matter what they or their former colleagues say to the contrary.

The money and real estate were likely given to ensure the couples' buy-in to the espionage operation; the research assistants with deep ties to China were likely there to augment whatever illicit transfer of proprietary and dangerous research was going on (as well as to keep a close eye on the husband–wife duo); the exfiltration of sensitive biotech data from the NML was limited primarily to Ebola, probably for perception management purposes. But the fact is that we will never know the full extent of this matter. Like so much in the biotech sector, especially as it relates to COVID-19 and the Wuhan Institute of Virology, the truth has been shrouded.

CHAPTER 11
TOM COTTON DEMANDS ANSWERS ABOUT COVID-19

At the height of the pandemic, multiple important personages – including United States senators who had access to classified intelligence reports – were convinced that COVID-19 was artificially made. Notably, US Senator Tom Cotton (R-AK) led the public in demanding that China come clean about the origins of the disease. Cotton, like anyone else who has dared to question its origins, was lampooned in the media.

Cotton is a former member of the United States Army who served two tours in Iraq, holds a J.D. from Harvard Law School, and serves on the Senate Intelligence as well as the Senate Armed Services Committees.[1] He has, for years, been thought of as future presidential contender for the Republican Party.[2] Whatever one's opinions about Sen. Cotton's political views, the notion that he would squander both his hard-won reputation and his future political ambitions by making exaggerated or false claims about the origins of COVID-19 is absurd.

Writing in the *Wall Street Journal* in April 2020, Cotton explained that, as a member of the Intelligence Committee, he believed that "The evidence is circumstantial, to be sure, but it all points to the Wuhan labs [being the source of the outbreak]." Cotton also pointed out that "[Chinese] laboratories working to sequence the virus's genetic code were ordered to destroy their samples. The laboratory that first published the virus's genome was shut down."[3] These are not the actions of a regime that is

interested in uncovering the truth. Such actions, instead, indicate a pervasive cover-up. Tom Cotton is one of the few truth-
seekers in a position of power and influence.

The Intelligence Committee is one of the most powerful
committees in the United States Senate. Comprising fifteen
members, the bipartisan committee was created in 1976 "to
oversee and make continuing studies of the intelligence activities and programs of the United States Government." And while
all Senators have access to secret intelligence (because all are
required to vote on matters affecting intelligence), it is only the
fifteen members of the Intelligence Committee that "have
access to intelligence sources and methods, programs, and bud-
gets." The Senate Intelligence Committee upon which Senator
Cotton sits is so powerful that the president of the United States
may "restrict access to covert activities to only the Chairman
and Vice Chairman of the Committee."⁴ So, the president can't
just cut this committee out of the intelligence loop entirely, as
he can (and does) cut out ordinary members of congress.

Because Donald Trump was president and was a member
of Cotton's own party, it is unlikely that he restricted access to
sensitive intelligence about the Wuhan lab to only the leadership of the Senate Intelligence Committee. In other words, Cotton was *not* spreading misinformation, as the censors in
America's media were claiming at the time. He knew what was
really going on because he had access to up-to-date intel. Interestingly, April 2020, when Cotton's *Wall Street Journal* op-ed was
published, was around the same time as the speech for which I
was punished for having spoken about Ralph Baric's gain-of-
function tests on SHC014 and its similarities to COVID-19. Our
intelligence community was aware of the lab leak hypothesis –
and increasingly became convinced of its legitimacy.

Alas, the politically minded censors in America's media
could not leave well-enough alone. Everyone – especially high-
ranking members of the Senate Intelligence Committee – was
required to toe the line drawn by Big Pharma, Big Tech, and the
CCP. So, the *New York Times* published an op-ed attacking Sen.
Cotton for his claims about the origins of the virus.

Who did America's so-called "paper of record" choose to

refute Cotton's claims? None other than a Chinese molecular biologist named Yi Rao, a proud member of China's Communist Party.

Yi's *New York Times* article had to have been written by the CCP's Politburo, because it was rich in counterclaims, false equivalencies, and projection. The headline of Yi's op-ed asserts that, "My Relatives in Wuhan Survived. My Uncle in New York Did Not." Of course, neither he nor the *NYT*'s "fact checkers" bothered to point out that New York City's death toll was so much higher than many other parts of the United States because both New York's mayor at the time, Bill DeBlasio, and the governor of New York, the disgraced Andrew Cuomo, enforced strict lockdowns that many experts today believe led to more deaths.[5] And no one took the time to highlight how many more people died in Wuhan as compared to New York City from the pandemic. The Chinese government still hides the true numbers, but they were far higher than Yi Rao cared to admit.[6]

Yi continues his obscene propaganda piece by arguing that his beloved uncle who lived in New York, if he had stayed in China, "would have been saved."[7] Yi is a man who proudly claims to be the father of China's Thousand Talents Program, the mother of all high-tech espionage projects directed against the United States and its allies. But the Gray Lady's editors failed to disclose Yi's connections to the CCP.[8]

Of course, the claims that Yi made in his op-ed were absurd. That the *New York Times* either did not care or did not know about Yi's extensive ties to the Chinese Communist Party is even more dangerous. As Zachary Evans of *National Review* brilliantly pointed out, the Chinese scientist not only developed the country's Thousand Talents Program, but he also writes extensively for *Caixin*, a scientific blog of which the CCP has an approximately 40 percent ownership. Yi's disingenuous op-ed, and the *New York Times*'s willingness to print it, is further proof of how corrupted our system has become, and how much influence the Chinese Communist Party wields therein. Peter Schweizer has rightly referred to this phenomenon as "elite capture."[9]

By the present moment, with so much of the controversy,

the censors have been proven wrong. After a year of withstanding attacks made by "pundits, politicians, and activists in white lab coats," Cotton was vindicated.[10] Almost every one of Cotton's claims (as well as my own and those of a handful of others, such as Steven W. Mosher) has either been proven correct or is now being actively investigated.[11] As it turns out, not everyone challenging the "scientific consensus" and the preferred political narrative are crackpots. Time, as Winston Churchill once said, heals all wounds. And over time, the truth will be revealed.

Just as I noted in earlier chapters, a coterie of supposed medical experts, such as the head of the WHO himself, reversed course after a year of denying the lab leak hypothesis. These former skeptics and censors now say that lab leak is our most probable explanation. Less than a year after President Biden ordered the shuttering of a State Department task force started by former President Trump to investigate the origins of COVID-19, the White House ordered the intelligence community to "redouble" its efforts to determine the origins of the virus.[12] Even the loudest skeptics are changing their minds. Naturally, these people expect the public to just nod along dutifully and not hold them accountable for their past idiocy. Instead, we are all to continue loathing people like Senator Cotton and other truth-tellers, who had the courage of their convictions to buck the "prevailing wisdom" of the time and challenge the narrative that the Communist Chinese (and their allies in the US government, the global scientific community, Big Tech, and Big Pharma) had crafted to keep us dumb and pliant.

But as you are about to read, Senator Cotton is not alone. There are others, mainly from the Republican Party, who are not going to let this matter go – no matter how many CCP agents the *New York Times* hires to write hit-pieces about them.

CHAPTER 12
RAND PAUL WANTS TO FIRE FAUCI
(WITH GOOD REASON)

The outbreak of the novel coronavirus from Wuhan, China, was the worst crisis of this young decade. As a proportional matter, it was likely the single worst event of the twenty-first century – outstripping the 9/11 attacks and the subsequent Global War on Terror, the Hurricane Katrina fiasco, and the Great Recession of 2008. The pandemic disrupted every aspect of our lives. Because of its biological nature, former President Donald Trump turned to the scientific community in the United States for help leading us through the crisis. Organizations like the Centers for Disease Control (CDC), the Food and Drug Administration (FDA), the National Institutes of Health (NIH) and its National Institute of Allergies and Infectious Diseases (NIAID), the Federal Emergency Management Agency (FEMA), and several other government bodies were all created precisely for the type of crisis the country underwent in 2020.

As citizens of a First World nation, Americans are inclined to trust medical professionals. During a pandemic, the men and women in the white lab coats are elevated to an almost godlike status in the eyes of the public. After all, these are highly educated people with decades of experience. Like a guild of priests in ancient times, the medical community speaks a language that few others understand, has access to exclusive knowledge, and can conjure seemingly magical powers to divine curatives for our ailments. In times of biological crisis, the medical community is essential. Their views and preferences can easily become the views and preferences of politicians who do not know any better.

For the most part, this is sound public policy. Pandemics are highly technical ordeals, requiring rapid and decisive responses informed by the facts. Scientists and medical doctors are naturally the ones to whom we'll defer. Yet this outlook for pandemic response is not a panacea. While "following the science" is key to understanding and defeating a pandemic, when a novel pathogen is ravaging the world, information is often scarce, and the science is *always* changing. You must be able to turn on a dime. That might be easy for an individual or a lab – but for a country of more than 300 million citizens? Not so much.

Once politics enters the equation, maneuverability is taken off the table. People, particularly partisans with agendas, become locked into strategies that may or may not represent the best course of action. What's more, scientific experts, despite their knowledge and experience, are not above reproach. They are, after all, only humans. And for those scientists who are employed by government agencies, such as the NIAID, following preordained political agenda becomes as important as following the science. And these scientist-bureaucrats use the shield of science's reputation to objectivity to hide their underlying, yet unmistakable, political agendas.

Dr. Anthony Fauci was brought on to advise former President Donald Trump at the start of the pandemic. He couldn't have come with a better resume. This was a man who spent decades at the NIAID, spearheading the US government's research into HIV/AIDS. A short, thin, bespectacled septuagenarian with a raspy Brooklyn accent, Fauci quickly became a news media favorite. Exuding quiet confidence, always looking intense, and speaking in a calm manner, Fauci gave off the air of a man who understood disease. He was the kind of general that you wanted to have leading your forces against an intractable foe.

Appearances are often deceiving. And in Washington, where Fauci spent the better part of his career, deception is the currency of the realm. Because of Dr. Fauci's decades of Washington experience and his role as head of the primary infectious disease arm of the US government, it was only natural that he would take the position of Trump's chief adviser on the pandemic. But Fauci was also among the most compromised of those scientists who could have been promoted to the role.

Recall that it was Fauci whose name was all over the NIAID and NIH grants given to the Wuhan Institute of Virology to conduct their risky gain-of-function tests. Riding into the White House as a man determined to save America – and the world – from the novel coronavirus, Fauci quickly began promoting inaccurate information. Early on, Fauci downplayed the whole thing – just as the leaders of China's Communist Party had done at the start of the outbreak in China.

While defenders of Fauci's early comments during the pandemic in 2020 say that he "qualified" his remarks to multiple television and radio show hosts, the fact remains that the point man for pandemic response in the United States government was telling the public not to worry about it.[1] He was saying this, by the way, as President Trump was taking steps to mitigate the spread of the disease by shutting down direct, public air travel between the United States and Asia (and eventually with Europe). For these actions the former president received significant pushback from medical experts and politicians alike.[2] Many of these critics would go on to make policy recommendations that ran counter to their beliefs and statements at the start of the pandemic.

On January 21, 2020, as the pandemic was just getting underway, Dr. Fauci went on the conservative *Newsmax* network and explicitly told host Greg Kelly that "[COVID-19] is not a major threat to the people of the United States, and this is not something that the citizens of the United States right now should be worried about."[3] Shortly thereafter, Fauci appeared on the Trump-friendly John Catsimatidis radio program and argued that COVID-19 wasn't "something the American people need to worry about or be frightened about."[4]

Interestingly, on January 23, 2020, Fauci expressly opposed COVID-19 lockdowns. Speaking on a podcast for the American

Medical Association, Fauci said that "There's no chance in the world that we could do [lockdowns] to Chicago or to New York or to San Francisco."[5] My favorite quote comes from a March 9, 2020, interview in which a member of the press asked Fauci if he believed political campaign rallies for the 2020 presidential election cycle should have been stopped until the disease was dealt with. Fauci explicitly said that "If you're talking about a campaign rally tomorrow, in a place where there is no community spread, I think the judgment to have it might be a good judgment."[6]

Many political leaders took Fauci's words to heart. As late as April 2020, several prominent Democratic Party leaders were insisting that the Trump Administration was overreacting to the threat of COVID-19.[7] Democratic Party Speaker of the House Nancy Pelosi of California toured Chinatown in her home district of San Francisco to prove that the Trump Administration was overreacting to the threat of the disease.[8] Then–New York City Mayor Bill DeBlasio encouraged his fellow New Yorkers to discard what he believed was the fearmongering of Trump Administration officials and "to make some dinner plans, do some shopping and stand with our neighbors!"[9] New York City, of course, would go on to have one of the highest death counts of any American city during the pandemic. Meanwhile, Ron Klain, who was serving as a senior Biden campaign official (and was eventually made Biden's White House chief of staff), was publicly sharing his skepticism about the severity of the novel coronavirus from Wuhan, China.[10]

These Democratic Party leaders were protesting the travel ban that Trump enacted to slow the spread of COVID-19 from Asia into the United States. This ban would inevitably be expanded to Europe as well.[11] They accused Trump and his team of engaging in anti-Asian rhetoric and making policy born out of their inherent racism – all because leading medical officials, like Dr. Fauci, were downplaying the severity of the illness.[12]

During the early part of the pandemic, Fauci and the medical response team for the White House urged Americans not to purchase personal protective equipment (PPE) unless they were directly dealing with people infected with the disease. The argument from the government was that one wouldn't ordinarily purchase PPE during flu season, for example. Why would Americans take such extraordinary steps to combat COVID-19 if it was in fact "nothing to worry about," as Fauci initially claimed? Of course, COVID-19 *was* something that should have worried every American. What's more, it would later be revealed that Dr. Fauci and other key members of the White House coronavirus task force (including Vice President Mike Pence, who was put in charge of the coronavirus response for the White House by President Trump) were lying about the need for Americans purchase PPE at the start of the pandemic. That was done to prevent America's limited supply from being drained by non-essential personnel.[13]

Government leaders anticipated the likelihood that PPE would be in high demand. In fact, by June of 2020, the country's PPE was depleted.[14] Since the United States imported much of its PPE from China, which had been shut down due to the pandemic and was hoarding a massive amount of PPE, the White House wanted to conserve as much of America's existing stock of PPE and keep it for medical professionals, government leaders, and other essential people.[15]

It was an understandable goal. But this bald-faced lie intended to manipulate the American people into *not* taking disease mitigation countermeasures before the pandemic got fully underway in the country should be troubling to everyone. Plus, if Fauci and his colleagues were lying about these things; if Fauci was out there downplaying the disease at the start of the outbreak in America for political purposes, what else was he lying about? And what political ends did those lies serve?

Regardless of what was said at the start of the pandemic, the country would ultimately endure caustic lockdowns, extreme social distancing measures, forced masking, and several other infringements upon our personal liberty that were done to "slow the spread." Some of it, while painful, may have been necessary. Much of it was not.

In any event, the methods of putting these actions into place were themselves unnecessary and divisive. And those methods, in many cases, derived from policies put forward by

Dr. Fauci and other government scientists who were wrong about much of the disease. Inevitably, the costs to the population in the form of economic hardship, mental health damage, stunted social and academic development of our young people, and overall harm done to our institutions were never considered by those who advocated for the harsh COVID-19 countermeasures. Nor will we know the true impact of these horrible things on our society for many years to come. But there were *some* leaders in the Senate, such as Rand Paul (R-KY), who immediately opposed the draconian countermeasures. Paul soon began questioning Dr. Anthony Fauci in his public hearings on the matter.

As a result of his constant conflict with Dr. Fauci, several truths that were otherwise hidden from the public's view about Fauci, the NIAID, the NIH, and the creepy closeness between America's scientific community and China's were revealed. Fauci does not come off looking very good. Even those who proudly planted "We Love Fauci" signs in the front yards of their brick mansions in posh Bethesda, Maryland, have begun to look unkindly at the government doctor.[16] After all, here is a man who was categorically wrong in almost all his policy suggestions (whether due to incompetence or deceit), from the very start of the pandemic in January 2020. But he has yet to be held truly accountable for these actions.

Senator Paul, an ophthalmologist by training and a Libertarian by birth, does not take infringements of liberty or abuse by medical practitioners kindly. One of the points of contention between Senator Paul and Dr. Fauci has been over the origins of the novel coronavirus.[17] Repeatedly, Fauci has claimed that there was "no evidence" that the disease originated from a lab in China. Yet while under oath in front of the Senate, Fauci insisted that he welcomed any investigation into its origins. This mismatch between Fauci's emphatic statements to the American people and his statements to political leaders is reminiscent of his dissembling the usefulness of PPE to the public. He sowed confusion when clarity was needed.

Fauci has an uncanny ability to take both sides of the same issue, which makes his advice suspect. The Left's Fauci Fan Club insists that Dr. Fauci is just being his notoriously cautious self.

These supporters argue that, in the face of unfair Republican criticism, Fauci was forced to choose "between the façade of nonpartisanship on one hand and professional integrity on the other, and he has clearly chosen the latter."[18]

There may, however, be something deeper here. Fauci has served every president from Reagan through Biden as NIAID director. He began his career in government way back in the 1960s. Fauci is a creature of Washington, D.C., and can deftly navigate its seemingly impossible political currents and bureaucratic eddies. He is clearly an expert in CYA. Taking both sides of a position gives you the ability to point at past statements to defend a current position; to always be right, all the time – and therefore, to secure one's job.

Fauci is given far too much credit for having served Republican and Democratic Party presidents alike. He has been at the center of the government's response to two major disease outbreaks: the HIV/AIDS epidemic during the Reagan Administration, and the COVID-19 pandemic during the Trump and Biden Administrations. Both examples have highlighted the incompetence of Fauci's crisis management. Just as with his response to the COVID-19 pandemic, Fauci initially responded to the HIV epidemic in the 1980s with skepticism. He downplayed HIV in the early months and years of that biological crisis. During that time, Fauci was described as a "murderer" by Larry Kramer, the famous gay rights activist, for his "inexperience" and "refusal to hear the screams of AIDS activists early in the crisis" which Kramer believed "resulted in the deaths of thousands of queers."[19]

Kramer and Fauci eventually developed a close friendship after years of sparring, but the criticism of Kramer and other prominent HIV/AIDS activists rings as true today as it did in the 1980s. It was Fauci who downplayed the severity of the illness – and helped to ensure that the Reagan Administration's response was slow. Though it has been well documented that the entirety of the Reagan Administration did not take the illness seriously, Dr. Fauci certainly did not help them to reverse their views on the matter.

In the past year, Fauci and Senator Paul have sparred over a range of issues related to the pandemic. They've disagreed about the lockdowns, about the efficacy and necessity of vaccine mandates, and crucially, about the origins of COVID-19. At every turn, the media has sided with Fauci. Yet Fauci has a long track record of being on both sides of whatever issue the two men were arguing over. Regarding the origins of COVID-19, one exchange stands out. In June of 2021, Fauci and Paul got into a contentious debate over the definition of gain-of-function tests and the origins of the COVID-19 virus. Senator Paul introduced an academic article into the record in which both the NIH grant number, along with Dr. Anthony Fauci, are credited for having provided the funding necessary for Dr. Shi Zhengli (the "Bat Woman of Wuhan") to conduct gain-of-function tests to take coronaviruses found in Horseshoe Bats and make them transmissible to humans in a lab setting.[20]

Squirming in his seat as Senator Paul read this fact into the public record during the Senate hearing, Fauci was quick to insist that the Senator, himself a medical doctor, was "egregiously incorrect" to claim that the gain-of-function testing that the NIH had funded in Wuhan had anything to do with the creation of COVID-19.[21] Fauci then attempted a convoluted explanation of how the research his organization helped to fund in Wuhan was *not* related to the COVID-19 pathogen. Of course, this is not what either Senator Paul or people like me are claiming. What we believe is that gain-of-function research on coronaviruses derived from bats was being funded by the NIH and related American organizations, and that research may have contributed to the creation of COVID-19 in the WIV … and that the data pertaining to the original creation of that lab-made pathogen has either been erased or shielded by the CCP.

Fauci obfuscated and deflected to the best of his ability, knowing full well that Paul was onto something. The reason that Fauci was so visibly uncomfortable, why he did his best to deflect attention from that which Paul had entered into the public record, was because just a few weeks prior to that Senate hearing, Fauci had explicitly denied that the NIH had funded any gain-of-function research whatsoever.[22] Senator Paul was calling Fauci out about the obvious lies he had told both to the public and to the United States Senate. Fauci then doubled down on his previous lies, in his tense argument with Senator Paul. Fauci had the gall to insist that he was neither disavowing nor denying that which he had said previously under oath, and that he would not retract any previous statements made about the NIH not funding coronavirus gain-of-function tests at the WIV.

Fauci was lying about funding gain-of-function tests for years at the WIV, even as the Obama Administration had imposed a strict moratorium on all gain-of-function tests in the United States as far back as 2012. In 2022, Fauci and Paul would again engage in a fiery debate over a related issue that may point to the motive for why Fauci and the NIH were so gung-ho about evading the Obama era moratorium on gain-of-function research in the United States. That motive was simple, and it can be uncovered by simply following the money.

Senator Paul is skeptical of the need for a COVID-19 vaccine mandate. He and other Republicans are concerned about the prospects of more or less forcing everyone in the country to receive vaccines that are, by definition, still experimental. As you've seen, the type of mRNA methodology used for the creation of the vaccines has not been tried in a major vaccine before (the process has been tried for rare illnesses such as Zika and Rabies).[23] It should be noted, however, that studies thus far indicate that those who've received the vaccine, even if they have gotten sick with COVID-19 since being vaccinated, generally do not suffer as badly as those who did not receive the vaccine.[24] At the same time, because the vaccines are little more than a year old, any studies of potential long-term side-effects have yet to be completed.

Therefore, a degree of skepticism should not be dismissed outright. Further, that the vaccines in the United States were developed under emergency protocols meant that certain legal protections were carved out for the pharmaceutical companies developing them. These protections were enacted to ensure that Big Pharma was quick to produce and offer the vaccines to a frightened public. But because of those protections, should long-term side effects be discovered, it's unlikely that these companies would have to pay damages.[25]

Speaking to the Senate by remote communications, Dr.

Fauci was confronted by Senator Paul about the issue of pharmaceutical companies paying out generous royalties to scientists employed by the NIH for their work on unique science experiments. The NIH and its cavalcade of supporting agencies, like the National Institute of Allergies and Infectious Diseases, "doles out roughly 32 billion worth of tax dollars annually in the form of research grants to pharmaceutical companies and the healthcare community," reports Adam Andrzejewski, the founder of a website dedicated to exposing government waste, *OpenTheBooks.com*.[26] "But [scientists at the NIH] also receive a hidden stream of private royalty payments for their innovations," Andrzejewski concludes. According to Andrzejewski, "roughly 1,800 scientists [at the NIH] have received an estimated $350–400 million in these payments, from entities like pharmaceutical companies, during the last decade."[27]

In fact, as Andrzejewski divulges in his report, when his organization sued the NIH to release details about these financial transactions, the "NIH ignored and denied lawful open records requests." The NIH was ultimately forced to disclose their records of these royalty transactions, but the government health agency – which American citizens fund with their tax dollars – is now "redacting the amount of each individual payment, and which company paid it." As Andrzejewski states, "Each royalty check is a potential conflict of interest."[28] As this book has argued, the connections among Big Government, Big Pharma, Big Tech, and the CCP are threats to ordinary Americans – and reports of the kind that *OpenTheBooks* has compiled further illustrate these dangers.

Senator Rand Paul had picked up this important thread in the summer of 2022. He and other Republican leaders such as Senators John Molenaar of Michigan, Josh Hawley of Missouri, Rick Scott of Florida, Jim Lankford of Oklahoma, and Ron Johnson of Wisconsin demanded answers about the royalty question from the Biden Administration's Acting Director of the NIH, Lawrence Tabak. It was Tabak who conceded that there was a conflict of interest and vowed to resolve the matter. Unmoved by the empty promises of the Biden Administration official,

Senator Rand Paul pressed Dr. Fauci. Another uncomfortable,
hostile exchange followed:

"Can you tell me that you have not received a royalty from any entity that you ever oversaw the distribution of money from research grants?" Senator Paul inquired of Dr. Fauci.[29]

Shifting in his seat, Dr. Fauci burbled, "Um, well first of all, let's talk about royalties – "

Paul was determined to keep the doctor on point: "No, that's the question! That's the question," he asserted in an exasperated tone. "Have you ever received a royalty payment from a company that you later oversaw money going to that company?" Paul asked again.[30]

Clearly being taken aback by Paul's incisive questioning (after all, for years, scientists from the NIH have been avoiding such tough questions from a succession of Congresses that simply did not utilize their oversight fully), Fauci answered, "You know, I don't know as a fact, but I doubt it."[31] It's possible that Dr. Fauci does not know how much money he has personally received from royalties after having been in government for decades. If that is truly the case, it could be because the amount has been so little that it's negligible to his personal wealth. Or, it could be so great that he feels it's necessary to obfuscate this truth to avoid being viewed in a negative light. It's likely the latter, as you'll see, given how long Fauci was in government and how many medical experiments his lab oversaw. Fauci is keenly aware that revelations about his (and other high-ranking government scientists') making a small fortune from royalties would be ruinous to his and the NIH's credibility.

Unsatisfied, Senator Paul continued with his questioning, "Well why don't you let [the Senate] know? Why don't you reveal how much you've gotten and from what entities? The NIH refuses – we ask [the NIH], we ask them – we ask them whether or not who got it, how much can you tell us? They [send the information] redacted! Here's what I want to know. It's not just about you. Everybody on the vaccine committee, have any of them ever received money from the people who make vaccines? Can you tell me that? Can you tell me if anybody on the vaccine approval committee ever received any money from people who make the vaccines?"[32]

At this point in their conversation, Fauci grew angry. "Soundbite number one? Are you gonna [sic] let me answer a question?"[33] Fauci asked. In fact, Senator Paul had given the NIAID director ample time to answer what should have been simple questions. But instead of answering, Dr. Fauci evaded. Fauci, obviously having had enough time to formulate a too-cute-by-half response to what was clearly an unanticipated line of questioning, started, "Okay! First of all, according to the regulations, people who receive royalties are not required to divulge them even on their financial statement, according to the Bayh-Dole Act."[34]

That was Fauci's first complete answer. Rather than immediately shoot down Paul's insinuation that the COVID-19 vaccine requirements had been concocted by government scientists who were on the take from Big Pharma, who benefited financially from getting as many people as possible jabbed with a vaccine that may or may not be effective, Fauci engaged in classic ass-covering behavior: he cited a Congressional act crafted in the 1980s by former Senators Evan Bayh (D-IN) and Bob Dole (R-KS) that effectively allowed for NIH scientists to hide their financial ties with Big Pharma.

Now, Fauci, who was clearly in the hotseat and was vamping for any face-saving bit of information he (and his staff) could quickly conjure up to end Senator Paul's unwanted line of inquiry, suddenly revealed this bit of data, "So, let me give you some example, from 2015–2020, I – the only royalties I have – was my lab and I made a monoclonal antibody for use *in vitro* re-agent that had nothing to do with patients. And, during that period of time, my royalties ranged from $21 a year to $7,700 dollars per year and the average per year was $191.46."[35]

Well, that certainly showed Rand Paul and all of us right-wing lunatics, right? Not quite. This is especially true since the NIH has been proven to withhold the specific data on which of their scientists receive royalties and how much (it took a lawsuit to get them to publicize even a fraction of this information).[36]

For his part, Fauci insists that he gives his royalties to charity. But no information has been found corroborating his claims.[37]

Fauci gave a bizarre, low-ball example to buy himself and his staff time to figure out how they can cover up more uncomfortable bits of data that are, like a poorly run Chinese BSL-4, leaking catastrophically damaging things into the wider world. Throughout his contretemps with Senator Paul, Dr. Fauci's phone was ringing – indicating that someone was feverishly sending Fauci texts in real-time. Fauci has run the NIAID since the Reagan Administration. He has worked as a government scientist since 1969. This guy had his gloved hands in multiple experiments – and you can bet your bottom dollar that he was making bank off the multiple research projects he was engaged in all these decades!

And just what was the "monoclonal antibody *in vitro*" project he was disclosing? It was an attempt to use monoclonal antibodies – the same technique that Xiangguo Qiu at the Winnipeg lab used effectively to cure Ebola in 2014–16 – to prevent the "acquisition of HIV infection and suppress HIV in infected individuals in vivo." As per the scientific paper that Dr. Fauci and several other scientists wrote on this matter in 2014, "Clinical trials have been planned and/or are underway."[38] It's likely Fauci was making more royalties as clinical trials progressed.

Fauci has been working for decades on finding a cure for HIV. His commitment to this cause, after having initially downplayed the disease, is admirable. HIV/AIDS is one of the most pernicious illnesses in our lifetimes; it is unique among more conventional illnesses in that it destroys the immune system. Yet Fauci's pursuit of this cure has become his White Whale. He has subordinated all other considerations to this goal. Inventing a vaccine for HIV/AIDS would ensure that Anthony Fauci is a name that history would not soon forget; it would also make Dr. Fauci a very wealthy man (and, likely, a Nobel Prize recipient).

Relatedly, there is extensive evidence pointing to the fact that the NIH, the NIAID, and the Wuhan Institute of Virology were collaborating on revolutionary HIV treatments. The 2019 annual report of the University of Maryland's Institute of

Human Virology elaborates that a scientist visiting from the Wuhan Institute of Virology, Wei Wang, was leading a study on how to treat "infections that cause pathogenic levels of inflammation."[39] In this same report, Drs. Anthony Fauci and Tae-Wook Chun, the two men who pioneered the use of monoclonal antibodies *in vitro* for HIV in 2014, are credited for their collaboration with the University of Maryland.[40] All these collaborations were part of the same project to address different aspects of the HIV/AIDS epidemic.

There's even one theory that the Wuhan lab–created COVID-19 pathogen was part of a larger, longer-running experiment overseen by Dr. Fauci and his team at the NIAID, along with Dr. Shi Zhengli of the WIV, to create a cure for HIV. When I initially heard of this theory, I rolled my eyes. Yet soon I happened upon a Swiss scientific research paper from 2006 that reads, "The molecular biology of coronaviruses and particular features of the human coronavirus 229E (HCoV 229E) indicate that HCoV 229E-based vaccine vectors can become a new class of highly efficient vaccines."[41] In April 2022, Dr. Fauci explained in a podcast how his NIH was supporting three projects, along with Big Pharma companies IAVI and Moderna, in Phase I trials that used the mRNA method to create a potential HIV cure.[42]

And remember Xiangguo Qiu and her HIV specialist husband, Keding Cheng, who were fired from Canada's only BSL-4 lab in Winnipeg? It turns out they were also looking at utilizing her Ebola research to fight HIV. They authored (along with a handful of other researchers) an article in May 2019 entitled "Incorporation of Ebola Glycoprotein into HIV Targeting Particles Facilitates Dendritic Cell and Macrophage Targeting and Enhances HIV-Specific Immune Responses." The article calls for the creation of "Dendritic cell (DC)-based HIV immunotherapeutic vaccine" and "macrophage targeting" to "induce a more potent immune response" against HIV infection.[43]

Given how many times our medical leadership has lied to us about everything surrounding COVID-19, is it possible that the pandemic was somehow connected to ambitious HIV vaccine research, of which Dr. Fauci and his team were integral parts? The people require firm answers on these matters, if the institutional legitimacy of our necessary medical organizations is to be maintained.

Senator Paul's continued grilling of Dr. Fauci, therefore, was justified. As you've seen, Fauci and his fellow scientists have contributed to the chaos of the pandemic by not being open with the public. From the very outset, Fauci and his fellow scientists in the West were happy to accept the CCP's obscene lies about the origins of COVID-19. They then misled the American people and the political leadership of the United States at the start of the pandemic in America about the severity of the illness and the proper countermeasures needed to stave off infection.

These same "experts" lied – and got the former vice-president to lie for them, as well – about the need for Americans to purchase PPE *before* the pandemic got out of control. Fauci and his cohort of "trusted" scientists got the lockdowns wrong, if not in theory, then certainly in action. After that, this same group censured anyone, expert and non-expert alike, who questioned the efficacy of COVID countermeasures, including the vaccines. In fact, they went beyond that, demanding that the political leadership impose harsh vaccine mandates on the American people.

As Senator Rand Paul and several other courageous GOP leaders have highlighted, Dr. Fauci and others at the NIH were, for years, reaping profits from Big Pharma on work they had been doing at NIH labs. In essence, there were financial and careerist incentives for scientists like Dr. Fauci to demand extreme, rapid vaccine mandates for the American people. Fauci and his fellow scientists decided to test mRNA vaccines on the largest sample group imaginable – us.

It is only now, in the shadow of the worst pandemic in decades, that the Senate is finally exercising its oversight power on an NIH that has become too comfortable with fraternizing with the very industry it is supposed to be overseeing (as well as with hostile foreign actors such as the CCP). Now that leaders like Senator Paul are demanding answers on all fronts, things are getting ugly. If the NIH's leaders give in and share with Congress the specific details of their collaborations with Big Pharma and hostile foreign regimes, heads will roll. If they refuse, whatever remaining public credibility the NIH has will be irrevocably lost.

CHAPTER 12

Demanding Answers

Once the GOP is back in control of the Legislative Branch, Senators Cotton and Paul (and others) should immediately lead an
investigation into the origins of COVID-19. As future chapters
will show, the links among Western scientists, biotech investors,
and the CCP are, even now, deepening. Far from being a unique
event in this century, given how deeply enmeshed the biotech
sectors of China and the West are becoming, it is likely that
COVID-19 could be the first of many manmade biological crises.
A wide-ranging investigation – not only into COVID-19's origins
and the WIV, but also into just what American money is funding in China – will be essential. And it will be essential to begin
this investigation soon, before it's too late.

Most Westerners believe that science will save us. It's a belief backed by a long history of scientific progress. When sicknesses such as polio plagued the world, it was the medical sciences that saved our species. There are few institutions that Americans still trust. But Americans tend to listen to their personal physicians, as well as the men and women in lab coats. Unfortunately, as in any institution, whatever altruism there may be, there is also greed, ego, and political ideology. This isn't new. Going back to the Second World War, when scientists basically ensured the United States would defeat the Japanese Empire with the development of the atomic bomb, the scientists who created those weapons had wildly different worldviews than what most Americans had.

Albert Einstein, for example, was a committed socialist.[1] J. Robert Oppenheimer, the true father of the atomic bomb, had longstanding open ties with the Communist Party of the USA.[2] Julius and Ethel Rosenberg, who had worked on the Manhattan Project, were integral members of a suspected communist cabal of scientists who believed that Washington could not be trusted as the sole keeper of the atomic bomb. Julius Rosenberg conspired to hand off atomic technology to the Soviet Union. The Rosenbergs' personal politics had encouraged them to conduct one of the most egregious acts of scientific espionage against the United States in modern history. Thanks to the Rosenberg handoff of American nuclear technology, along with the sharing of other atomic secrets with the Soviets from other prominent American and British scientists who were loyal to the communist cause, the Soviets were able rapidly to progress their indigenous nuclear weapons program. This, in turn, ensured a Cold War that would cause nearly three generations to live under the constant fear of nuclear annihilation.

For their crimes, the Rosenbergs were arrested, convicted as spies, and sentenced to death. The question that so many people ask is, "why did they do it?" Weren't these scientists well provided for? Of course, they were. These scientists had all the material comforts that should make someone enamored with the American way of life. But these individuals were nevertheless deeply dissatisfied with their country.

It is believed that the origins of the spy ring in which Julius Rosenberg played a decisive role resided with Klaus Fuchs, the notorious atomic scientist and communist spy. Fuchs had been a proud member of the Communist Party in interwar Germany. When the Nazis rose to power, he knew that he'd be a target of their political violence, so he fled to Great Britain. He ultimately became a British citizen and continued his cutting-edge work on atomic bomb research alongside notable British physicists, Nevill Mott and Max Born.

Soon, Fuchs was folded into Britain's atomic weapons program by Rudolf Peierls. But Fuchs could not shake his commitment to the communist cause; he believed that his scientific efforts (and those of other scientists) could contribute to the communists' utopian goals. So, Fuchs betrayed the country that had taken him in as a political refugee. As early as 1943, Fuchs began passing atomic secrets to Soviet spies. Once the Brits began working with the sprawling American "Manhattan Project," he began sharing that information with the Soviets, too.[3]

Fuchs was soon able to recruit scientists on both sides of the Atlantic to his cause. He fathered a massive, transatlantic, pro-Soviet spy ring of elite British and American scientists who willingly handed over advanced secrets to the Soviet Union. In the eyes of these scientists, they were saving humanity by depriving an inherently unfair capitalist country, the United

States, of an immoral monopoly on the atomic bomb. If atomic science were left exclusively in the hands of the capitalists, these scientists believed, humanity would be held under the jackboots of inequality and poverty.

Fuchs was close with an American military scientist named David Greenglass. Greenglass's sister was none other than Ethel Rosenberg. Ultimately, Greenglass wooed his brother-in-law, Julius, to the communist cause. According to one *New York Times* report, Greenglass was preaching communism "Everywhere – even at Los Alamos" where he "tried to persuade American G.I.s and co-workers that they would someday prosper in a utopian society free of squalor and injustice."[4]

This was the siren song of communism. It promised utopia but delivered the very things it strove to eradicate – squalor and injustice. These scientists were brilliant theoreticians, but they did not understand the real world and the practicalities of human nature. For them, the world was a giant engine to be tweaked and manipulated according to their expert designs – just like the technology they were innovating in the lab. But the world is not a lab.

Sadly, these great minds, led astray by wishful thinking, "believed in the USSR, dreamed of a Socialist state without discrimination.... Little did they know about Stalin's bloody purges and the Gulag," as the Russo-Israeli historian Zakhar Gelman famously wrote.[5] For this vicious system of government, many of America's best and brightest were willing to risk their lives and reputations. The Western scientists were dupes. But these "useful idiots" would have profoundly negative impacts on Western societies because of their naivete.

Great Minds Susceptible to Bad Ideas

Beginning in the late 1700s, as the clerical scholars were declining, new secular philosophers were rising to take their place. In the wake of the Enlightenment, in which so many had turned on the Christian thinkers and ideas, these new secular thinkers believed they could reshape Western Civilization and humanity to their liking. The brilliant Cambridge historian Paul Johnson looked at a handful of these secular thinkers in his excellent 1988 book, *Intellectuals: From Marx and Tolstoy to Sartre and Chomsky.*

In this magnum opus, Johnson identifies how secular intellectuals from the late 1700s to today believed that they alone could dictate how Mankind should order its society and what Man should believe. Whereas the previous intellectual elite, the clerics, based their legitimacy on their shared faith, these secularists had no common faith. Instead, as Johnson assesses, these philosophers insisted that their pursuit of moral and intellectual truth (as they saw it) qualified them for enjoying such a vaunted status in society. Yet as Johnson observes, none of his subjects actually told the truth. In the case of Karl Marx, the father of communism, he never had any significant relationship or contact with the working class that he claimed to represent in his writings. Instead, Marx and other intellectuals preferred to use abstract theories and concepts, rather than real people, as the basis of their thinking.

This same mentality was held by many of the scientists who partook in the great betrayal of America during the Manhattan Project. And this same mentality can explain why people like Dr. Fauci and others have no compunction about doing highly dangerous biological experiments with the Communist Chinese. Or why they were demanding the imposition of highly damaging lockdowns and other pandemic countermeasures, no matter the risks involved. Perhaps they're not ideological fellow travelers with the CCP. But the ego of today's scientists likely leads them to believe that the politics of the CCP are irrelevant. What matters to them is pursuing research as quickly as possible.

As Paul Johnson explains in his work, the egoistical nature of these minds, along with the fact that they know little outside of their area of expertise, makes them dangerous. If their more ambitious and arrogant ideas are not contained within their respective fields, society will suffer. Scientists working to cure illnesses have cut corners. They have obfuscated the truth about the origins of COVID-19. What's more, they have shared highly dangerous technologies and techniques with the CCP – which most certainly will incorporate these into China's growing war machine – with little care or regard for the safety of the American people.

Think back to the quote in the previous chapter in which Trump Administration officials joked how almost every government scientist voted against Donald Trump. Hardly anyone working at the NIH or its sister organizations agrees with roughly half the country's political preferences. When health policies are impacting all aspects of the public during a pandemic, it is a problem if those policies are being shaped by leftist ideology. The prevailing wisdom and political preferences of this group engender what is described as the only sensible or moral course of action. As you have read, however, the response to the pandemic has been completely untethered from reality. At every turn, the decisions were based less on the data and more on the politics of the scientists running the show.

Many questioned the validity of the national lockdowns, but these voices were overruled by the Faucis of the world. According to Pew Research, "in the first weeks of the [COVID] outbreak, [there was] bipartisan agreement on the necessity of travel restrictions, business closures."[6] Yet, those decisions were made based on limited data (much of which was incomplete and, when coming from China, unreliable). Over time, those opposing the lockdowns and draconian countermeasures should have been given a fairer shake. The arguments of those non-medical experts, like White House economic advisers Peter Navarro and Larry Kudlow, who were warning Dr. Anthony Fauci and his cabal about the economic disaster that prolonged national lockdowns would cause the country, should not have been ignored. Like so many in those opening weeks of the pandemic, I too supported the lockdowns. But once it became clear that the government had no plan and little understanding of the disease and how to combat it, I, like so many Americans, began questioning the efficacy of the countermeasures.

Most people believe that the novel coronavirus leaked out from a lab in Wuhan.[7] Our egghead class laughed at us and censored us for daring even to utter such notions. Turns out, it is likely true – as even the head of the WHO is now acknowledging. We have allowed scientific experts to have far too much sway in wider public policy debates; even having influence over matters related to censorship on the internet, which is both shocking and aggravating. That these experts have a track record of getting it wrong, are prone to overstatement, and, worse, have a history of consorting with enemies of the state should give every policymaker pause.

Follow the Scientism!

With the loss of formal religion in the West, we have replaced the religious dogma of our forefathers with the scientist dogma of today. Most people mindlessly defer to the men and women among us in lab coats – no matter how extreme their ideas may be. Whereas at the start of the century, most people had a healthy degree of skepticism about scientific endeavors, today they are browbeaten into accepting the latest scientific fad or theory. During the pandemic, as you've seen, they weren't just browbeaten, their digital personhoods were deleted by censors across social media platforms when they deviated from the prognostications of our expert medical class.

"Scientism" is an "excessive belief in the power of scientific knowledge and techniques."[8] This is the state of most of the Western scientific community, and it has shaped wider society – particularly as civic religion has precipitously declined in America.[9] According to a 2009 Pew Research Poll, most American scientists are overwhelmingly liberal.[10] As *Nature* reported in the aftermath of the 2020 presidential election, an astounding 86 percent of American scientists polled said they voted for Joe Biden for president.[11] This has prompted one writer at *Nature* to inquire if this lack of diversity among scientists' political opinion is a problem.[12] It explains why anyone who even questioned the official origin-story of COVID-19 was steamrolled by the majority of the scientific community. It also explains the abuses of most scientists relating to Big Pharma royalties awarded to NIH researchers. Few seemed to question the fact that this could represent a conflict-of-interest.

These eggheads have done untold damage to not only their

own industry and institutions, but our entire society. The pandemic eviscerated us and, at times, empowered the Communist Chinese. That the pandemic may have been created in a Chinese bioweapons lab that was partly funded by the NIH – and then covered up – is galling. What's more egregious is that the scientific community has clearly tried to prevent Congress from discovering these facts, as evidenced by, among other things, Dr. Fauci's evasive Senate testimony. Fauci and others act this way because they believe they are intellectually superior to the rest of us.

Unless Americans wake up to the dangers of scientism and the corruption in our scientific community, egregious events such as those that transpired during the COVID-19 pandemic will continue to occur. Accountability is a foundational element of a free society. In this case, "accountability" is just another word for "responsibility." Or, as the Reverend Robert R. Sirico wrote for the Acton Institute, "The end of freedom is, by necessity, the end of responsibility, and the end of responsibility is the end of civilization."[13] The scientific community embodies the cult of expertise that sacrifices the wellbeing of a country to the whims of a technocratic elite. Its members have arrogated to themselves the mantle of power and have, in turn, silenced anyone who dares to challenge them. Such power must be wrested back from their gloved fists and returned to the American people.

CHAPTER 14
THOUSAND TALENTS

When I began dating the woman who would become my wife, she had just gone through qualifying exams for her doctoral thesis at Yale University. As a third year Ph.D. student in Yale's genetics program, Ashley was part of the storied academic elite. Yet like so many students in America today, my wife was saddled with onerous debt from her long journey in our country's needlessly expensive educational system.

Even though she had not finished her studies, the mere fact that she was a Yalie gave my soon-to-be wife an enviable cachet among possible job recruiters. Not only were genetics research labs in Switzerland and Italy hoping to hire Ashley, there were countless biotech enterprises around the world looking to hire a woman with a sterling professional and academic pedigree. Many of these operated in places that, in 2014, may have sounded alien to a young woman from Virginia – places like Wuhan, for example. By the time of this writing, for all the wrong reasons, the ancient Chinese city of Wuhan has become a household name worldwide.

Chinese recruiters made an enticing appeal to students of America's top institutions, one that could be summarized as the following: "Come to China and begin your career in our state-of-the-art, freewheeling, leap-without-looking high-tech research and development sector. We'll pay you more in your first year than you'd ever make in the United States. And unlike in the

West, as a newly minted researcher from an Ivy League institution, *you* will be treated as the rockstar, not your boss. In fact, you could be your own boss by starting up a research lab on your own here in China."

No mention, of course, of the CCP's watchful eye. The pitch might continue: "You're still not interested in moving halfway across the world to a foreign nation ruled by a despotic regime? Well, how about, as a signing bonus, the Chinese firm [read: a front for China's government and military] pays off your entire student debt [in exchange for you working exclusively in China's high-tech R&D ecosystem]?"

Of course, since I was working on Capitol Hill and deeply ensconced in national security policy at that time, there was no way Ashley and I were ever going to move to China so that she could work for a CCP front organization. What's astounding, though, is how many of her graduate school cohort were enticed by the offer. Part of this was born out of desperation: they had to pay back their obscenely high student loans, and the quicker they did that the better.

Another part, though, was born of naivete. Many scientists are naïve about national security concerns. Why wouldn't they be? They spend much of their adult life in cloistered labs and classrooms pontificating about the esoteric; they belong to an elite, cloistered subculture searching for the next great innovation that will change all of our lives. Petty matters of national security are left to the simpletons like me. Another factor is their unchecked ambition. Who wouldn't want to jump to the front of the research line right out of school, to live the exciting life of a young expat in a totally alien culture, and to be handsomely rewarded for one's genius?

How many of America's next generation of talent are being targeted and wooed over to China, when otherwise they'd have stayed here in the United States and contributed to America's economy and system of innovation? This brain-drain will have deleterious long-term impacts on the future economic growth, as well as the national security, of the United States.

Unbeknownst to my future wife in 2014, she and her cohort were the targets of a sprawling and ambitious CCP project to bring as much of the world's scientific talent to China as possible. Known as the "Thousand Talents Program," the project was started in 2008. The renowned Chinese molecular biologist Yi Rao has been credited as being a key figure in its creation.[1] You'll remember that Yi Rao was the Chinese scientist who wrote that vicious *New York Times* op-ed attacking US Senator Tom Cotton for his stated skepticism about the official origin story of COVID-19. Other Chinese scientists and government leaders played significant roles in the creation of the Thousand Talents Program.

The Thousand Talents Program began as a "scheme to bring leading Chinese scientists, academics and entrepreneurs living abroad back to China."[3] China has always relied on its vast diaspora of expat scientists to supplement its ongoing quest to defeat the West in the race to dominate the fourth industrial revolution.[4] In fact, there are multiple examples in recent history of Beijing using a combination of patriotic appeal and financial incentives to induce the "cooperation" of Chinese scientists based in the Western world.[5]

But the CCP's talent acquisition program did not only target Chinese nationals. In 2011, "the scheme grew to encompass younger talent and foreign scientists, and a decade later, the Thousand Talents Program has attracted more than 7,000 people overall." Many of these scientists were of Chinese descent who felt a strong cultural and financial incentive to return to China. Still others saw an "opportunity to join the Chinese system with major administrative hurdles removed."[6]

As Christopher Burgess, a man who spent thirty years in America's Central Intelligence Agency, argued in 2021, "China's Thousand Talents Program is harvesting U.S. technology through the false promise of recognition and open sharing." Burgess goes on to assess that "[The Thousand Talents Program] is a cover to entice members of both U.S. academia and industry to share information which they would otherwise keep confidential with the sole purpose of advancing China's interests."[7] Burgess is not alone in this belief.

The Federal Bureau of Investigation (FBI) is so concerned with the program that it has created an entire website dedicated to defining the Thousand Talents Program and the threats it poses to the United States, as well as to our nation's companies and allies. Here, the FBI states that the Thousand Talents Program incentivizes "its members to steal foreign technologies needed to advance China's national, military, and economic goals." According to the FBI,

> Participants enter into a contract with a Chinese university or company – often affiliated with the Chinese government – that usually requires them to: subject themselves to China's laws, share new technology developments or breakthroughs with China (they can't share this information with their U.S. employer or host without special authorization from China), recruit other experts into the program – often their own colleagues.[8]

The FBI has classified this program as a significant and prolonged counterintelligence threat to the United States. The Bureau's website goes on to address how many American firms and research labs are completely unaware that they are being targeted by China for the purposes of high-tech, industrial espionage. Firms and innovators see dollar signs, while research labs see the opportunity to speed up the development of their research.

As for China, it sees in these gullible and greedy Westerners nothing but marks to be conned. Think of the Chinese attitude toward American researchers, labs, and biotech firms as similar to those of ambitious Europeans who came upon arable lands during the colonization of North America. When European colonists saw that the only thing stopping them from owning that rich land was the local native tribe, they took advantage of the natives. Those rapacious colonial Europeans tricked the native tribes into selling them that land for mere trinkets. These trinkets were shiny and meaningless, but they satisfied the natives for a moment – long enough for them to vacate their homeland. Only later, once it was too late, did they realize they'd been swindled.[9] This is how China treats Western scientists and research: as exploitable fools sitting on a veritable treasure trove, who can be conned into trading away that treasure for short-term glitz. All the while, China's capabilities improve enough for it to threaten America's dominant position in the world system and, ultimately, to found a new world order that favors the communist East over the democratic West.

Some reading this may find poetic justice in the United States being treated the same way America and her European progenitors treated the indigenous peoples of North America. But whatever sins you believe America may have committed as a nation, it is essential to understand that Americans have continually striven to uphold what they believe to be an international order that prizes freedom. This belief, for which so many Americans have fought and died for over the centuries, has built a world that is generally more open and prosperous than it has been at any other point. Allowing for China to displace the United States as the dominant world power would herald, instead, the birth of an order built entirely by totalitarian leaders determined to remake the world in the image of their own odious values. China's scientific–industrial espionage is meant to serve just this purpose.

The Thousand Talents Program is a brilliantly conceived and executed project in that it is a relatively low-cost way to achieve high, long-lasting reward for China. China gains access to top-tier American talent. It then learns proprietary trade and innovation secrets as part of these exchanges, and incorporates those secrets into its growing indigenous high-tech R&D sector. That sector, ultimately, is wielded to outperform American companies and labs and to threaten the US military's position in the Indo-Pacific and beyond. This is how China stays competitive with the United States.

In some cases, American firms and scientists have willingly signed onto the Thousand Talents Program – more on this later. In other cases, though, American firms, labs, and researchers are completely oblivious to China's ulterior motives. After all, as even the FBI website articulates, countless countries utilize talent programs as a means of jointly developing new innovations and enhancing their R&D ecosystems via cooperation. And China has deftly enmeshed their malign strategic goals within altruistic-seeming aspirations to conduct jointly developed scientific research.

This preference for mutually beneficial cooperative relationships explains why the leaders at Canada's only BSL-4 lab in Winnipeg did not demand answers about Xiangguo Qiu's requests to share highly sensitive Ebola and Nipah samples with the WIV until *after* the Royal Canadian Mounted Police counterintelligence investigators initiated an inquiry into the matter. For those researchers, science was borderless and global. It was only after Xiangguo had sent her samples without using the proper procedures that Canadian authorities had figured out what was happening and acted. By then, the damage was long since done. Yet China, unlike the Western democracies, is ruled by a nationalist regime intent on restoring what its leaders believe is China's rightful place as the "Middle Kingdom": the center of the world and the world's hegemonic power.

The push for scientific progress by any means necessary, the knowledge that researchers must "publish or perish," has created an American scientific community that takes irresponsible risks in its collaboration with China. As is usually the case with the United States, America's totalitarian enemies use our openness against us. Al Qaeda did this in the run-up to 9/11, when they infiltrated our porous immigration system with nineteen hijackers on student visas. Today, China uses our open research model and our scientific community's insistence on collaboration to further its own national interests.

The NIH awards countless grants per year to programs and researchers around the world. For years, this system has focused solely on the efficacy of the research and the results. Until very recently, no serious care was given to the prospect of foreign agents seeking access to NIH funding and proprietary research. According to security experts, "biomedical research espionage falls into the category of 'nontraditional espionage.'"[334] The US government has slowly but rightly grown attentive to foreign penetration of sensitive NIH databases via generous NIH grants. According to Joanne Carney, director of government relations at the American Association for the Advancement of Science, Washington has become concerned that American researchers may be creating "shadow labs [in China] working on the same projects as U.S. labs; failing to disclose funding and affiliations with other governments; and violating the peer review process."

On August 20, 2018, NIH director Dr. Francis Collins circulated a letter he had drafted to ten thousand NIH–affiliated facilities around the country. In the letter he addresses the budding topic of biomedical research security (or lack thereof) at the NIH. The letter – which should be seen by future historians as a pivotal document in which American leaders began sensing the threat of China's biotechnology espionage program, but were still too slow to recognize its extent and depth – argues that the "NIH is aware that some foreign entities have mounted systematic programs to influence NIH researchers and peer reviewers and to take advantage of the long tradition of trust, fairness, and excellence of NIH–supported research activities." Collins went on to advise his colleagues that these new security concerns will be "supported by a working group of the Advisory Committee to the (NIH) Director that will tap experts in academic research and security to develop robust methods" better to defend against high-tech biomedical espionage.[10]

One notable example of a Chinese "scientist" coming to the United States to conduct research for the United States – and receiving millions of taxpayer dollars from the NIH for research, only to happily share that research with China – is the Ohio State University's Dr. Song Guo Zheng. A biomedical researcher, Dr. Song was a key fixture at both his home university of Ohio State as well as Pennsylvania State University. Song was accused and ultimately found guilty of having falsified NIH grant applications to receive upwards of $4.1 million from the US taxpayer to fund cutting-edge research in the fields of rheumatology and immunology. In fact, it is believed that because of his NIH grant Dr. Song was given access to classified material from the US government in order to conduct his experiments. But Dr. Song ran into trouble when he decided to use both the classified information from US government databases and the $4.1 million he received from the NIH to "develop China's expertise in rheumatology and immunology."[1]

The FBI had been tracking Dr. Song for some time, since they realized he had lied on his NIH grant applications. The US government soon determined that Dr. Song was a Chinese national living in the United States on an expired passport. He also hid his affiliations with five research institutions in China. When Song Guo Zheng was arrested by the FBI on May 20, 2020, he was intercepted getting off a charter plane in Anchorage,

CHAPTER 15

Alaska. The biomedical researcher was attempting to board another chartered flight bound for China. Upon his arrest, the doctor was holding a briefcase carrying two laptops, three cell phones, several USB drives, silver bars, expired Chinese passports, and deeds for property in China.[2]

This behavior is less that of a biomedical researcher and more that of a spy hoping to evade capture in the midst of an exfiltration scheme. Song Guo Zheng likely had laptops and multiple USB drives containing an assortment of classified US government biomedical research that he had pilfered and was planning to give to his handlers in China's intelligence services upon arrival. He probably had multiple cell phones to coordinate his escape from the United States. The doctor had silver bars in case things went badly and he needed to get out of some sticky situation without using currency that could be tracked by US government officials. Song chartered a private plane for his escape and had it routed to Anchorage, a sleepy, provincial town where he might avoid detection and then board another private plane to China.

After his arrest, Assistant Attorney General for the Department of Justice's National Security Division John Demers remarked that "American research funding is provided by the American taxpayer for the benefit of American society – not as an illicit gift to the Chinese government."[3] Song Guo Zheng was sentenced to thirty-seven months in federal prison and was ordered to make financial restitution of $3.1 million to the NIH and $413,000 to his employer, the Ohio State University. Investigators determined that Song had been an integral member of China's Thousand Talents Program since 2013.[4]

Yet, if as I suspect, Song Guo Zheng was but one of many actors involved in a much deeper Chinese plot to gain access to classified US biomedical research, then the damage to America's economy is far more severe than US authorities are letting on. If the Chinese have spent at least a decade penetrating the NIH and thus US military organizations like DARPA, then it is very possible that the current revolution in China's high-tech industry is, in part, being fueled by massive amounts of US taxpayer–funded R&D whose theft the US government only recently detected, long after much of the damage was done. Further, it is possible that the US government, to save face, is now attempting to obscure the full extent of the damage that China did to the country.

Mr. Demers's argument that US taxpayer–funded science should not serve China was correct. But by 2018, when arrests similar to that of Dr. Song started happening, the Americans were far behind the curve. China had been running its Thousand Talents Program in the United States since 2008, exploiting American ignorance for a decade's time. It had coopted and snookered American academia, investors, and innovators in the biotech field – all to enhance its competitive edge, for both commercial gain and military advantage. Even after the outbreak of COVID-19, when it should have been clear that the risks far outweighed the benefits, American researchers and even some government officials still insisted that we must continue to collaborate with China on biotech R&D.

CHAPTER 16
CHINA STEALS US GOVERNMENT-FUNDED CANCER RESEARCH

Francis Collins's 2018 letter focused on the NIH grant program, and with good reason. In fact, the letter appears to have been prompted by a specific incident involving at least five cancer researchers who were affiliated with the MD Anderson Cancer Center (part of the University of Texas system) and who had received funding through an NIH grant. Of those five scientists, three were removed from their positions after an internal investigation revealed "serious" violations made by those individuals regarding confidentiality of peer review and the disclosure of foreign ties.

MD Anderson is endeavoring to create better treatments, and possibly a cure, for cancer. The organization receives 148 million in US taxpayer dollars from the NIH every year. One of the individuals who was placed under suspicion by authorities was charged with having violated peer-review confidentiality by emailing proprietary information to a colleague in China. Another accusation was that "several researchers had 'active and well-supported research programs in China'"; some had previously undisclosed financial ties to Chinese entities, and three were directly involved with China's Thousand Talents Program. The investigation was the conclusion of a much deeper look into biomedical espionage that the FBI had begun more than a year earlier.[1] Many of the details of that sweeping FBI counterintelligence investigation remain unknown to the public.

The MD Anderson Cancer Center is based in Houston, Texas. During the 2020 race riots that tore the United States apart in the aftermath of the George Floyd murder, the Chinese consulate was rumored to have housed a PLA intelligence unit that was providing financial support and assistance to violent protesters.[2] A deeper investigation into the Chinese Houston Consulate by US domestic security services later determined that the consulate in Houston was a proverbial beehive of espionage activity by China's intelligence services. The Trump Administration closed the Chinese Houston Consulate that summer. When asked to explain the Administration's reasoning, then–Secretary of State Mike Pompeo said that "[The CCP is] stealing … American intellectual property … costing hundreds of thousands of jobs."[3] It would be unsurprising to find that the suspected agents operating at MD Anderson had handlers assigned to them from the Houston consulate.

Shortly after the Houston consulate was shut down, there was a standoff between FBI officials and representatives of the Chinese consulate in San Francisco. At that time, a young Chinese woman, Juan Tang, who had enrolled at the University of California, Davis, and had entered the United States on a student visa, had fled American authorities, holing up inside the Chinese consulate.[4] US counterintelligence had called for her arrest after evidence surfaced that she had lied on her visa application about never having served in the PLA. As it turns out, Dr. Juan Tang did serve as an officer in the PLA Air Force (PLAAF). She had officially come to the United States to conduct cancer research at a UC Davis lab, but clearly the US government was sufficiently concerned about her work to chase her into the consulate.

Interestingly, the Department of Justice dropped all charges against her in July 2021. Attorneys for the DOJ reasoned that being convicted of lying on a visa application would have gotten Dr. Juan a maximum sentence of six months, "less than the ten months [she had] already spent in jail and under house arrest."[5] Still, one cannot help but to speculate as to why FBI counterintelligence did not squeeze her for more information.

CHAPTER 16

Biden's Cancer Moonshot Becomes China's Treasure Trove of Espionage

In 2016, President Barack Obama proudly announced his Administration's creation of a "cancer moonshot" program.[6] Taking its cues from President John F. Kennedy's original program to get American astronauts to the moon in a decade or less, Obama's "cancer moonshot" planned to "hit the gas" on oncology research over the course of five years. The NIH became heavily involved.[7] Obama placed his vice president, Joe Biden, in charge of the effort.

Lots of money and talent were flowing into cancer research during this time. Beginning also in 2016 was a massive surge in Chinese interest and investment in cancer research. In fact, that year China began testing highly experimental uses of CRISPR aimed at gene-editing cancer cells in order to render them benign. That same year, China was the first nation to test CRISPR-Cas-9 gene-editing technology on humans. Many in both countries viewed the Chinese "victory" over the Americans as a "Sputnik 2.0" moment.[8]

The Chinese scientists involved with the event reported their success in a peer-reviewed medical journal in 2020.[9] While the "outcomes were modest," the experiment was still more promising than most believed could be possible. In fact, one Chinese patient with late-stage lung cancer, a fifty-five-year-old woman, saw her "tumour [sic] mass initially [shrink], and her disease remained stable for almost seventy-five weeks before initially progressing."[10] The Chinese scientists who conducted the tests believe that using these experimental CRISPR-Cas-9 treatments in conjunction with other cancer treatments – at earlier stages in the disease – will eliminate the illness. It was a momentous victory for China in the battle for biotech supremacy with the United States.

It is believed that the competition for dominance in space, and the technologies that undergirded that important industry, created revolutionary new technologies that catapulted human development and helped the Americans win the Cold War. There appears to be a similar race occurring in biotech between the

United States and China. Curing cancer has been the first arena of battle, one in which China is emerging victorious.

Might we suppose that China was able to leapfrog the US by incorporating data compiled by scientists who were ostensibly working for MD Anderson and receiving copious NIH funding in the wake of the Biden-led "Cancer Moonshot" program? As noted previously, the FBI began a sweeping counterintelligence investigation more than a year before the scientists at MD Anderson were removed from their positions. Who's to say that the FBI was not tipped off to nefarious connections between American, NIH–funded cancer researchers and Chinese cancer research centers, or to Chinese spies operating in the United States on NIH grants?

China's invention of a potential, rudimentary cancer cure before the Americans could create one would give China the almighty first-mover advantage. While there are downsides to being the first company or country to develop new innovations (rivals can simply copy or make a better version of what you've created), the fact is that often getting one's product to the market first confers enormous wealth and competitive advantages over one's competitors. If it is true that China was able to develop paradigm-shifting cancer treatments with CRISPR-Cas-9 using research their spies pilfered from the Americans, then those enormous advantages will have gone to China rather than the United States, even though America was pioneering the research. What's more, such victories would also likely attract investment and talent to China that otherwise would have gone to the United States.

Further, the COVID-19 pathogen is believed to have, in addition to HIV and Malaria genes, cancer cross-sections in its genome. There are active studies that utilize cancer drugs to treat COVID-19.[11] Plus, the same research that likely supported China's experiments with gene-edited cancer cures also support research into mRNA-based COVID-19 vaccines. It's believed that mRNA vaccines could be the key to curing cancer, in addition to COVID-19 and HIV.[12]

There are many strange things happening in China's biotech sector today, and much of it stems from China infiltrating every layer of America's biotech R&D infrastructure. What's more, it is obvious that whatever exchange exists between the United States and the People's Republic of China, it is a one-way street: America's innovators and investors give to China, while China takes from America.

In fact, it's possible that the Obama Administration's "Cancer Moonshot" program was itself a reaction to China's unexpected leapfrogging of America in the critical domain of biotech. Too bad for the Yanks that they didn't think to seal up their leaky research pipes first. Finding a cure for cancer as quickly as possible is a good thing. I've lost friends to this illness and want nothing more than to see it defeated. Competition, far more than cooperation, often yields dynamic, high-tech solutions in a relatively short time. Yet the Chinese aren't just competing with the Americans. They're cheating. And the Americans are too naïve to recognize how dirty the Chinese are in waging this battle for biotech dominance. And, as always, Chinese medical research developed ostensibly for benign purposes is actually being folded into a far more dangerous and ambitious bioweapons program.

China stealing proprietary information from the United States – information gained through taxpayer funding – won't help Americans. These activities only empower China at America's expense.

What few in the public realize is that the United States sits atop a proverbial ecosystem of research and development. The NIH and its NIAID, as well as the National Cancer Institute (NCI), the National Human Genomic Research Institute (NHGRI), the National Science Foundation (NSF), and other alphabet soup agencies coexist and often share research and scientists on projects aimed at furthering the health and safety our nation, and the world. There's something beyond that, though. These agencies are linked, however indirectly, to the US military. They operate in tandem with US military organizations to further research goals.

Specifically, the NIH, as the flagship civilian research agency in the US government, shares personnel and resources with the Pentagon's Defense Advanced Research Projects Agency (DARPA). DARPA is a unique agency in that it is "responsible for the development of emerging technologies for use by the military." When you think of DARPA's inventions, think of the stealth plane, the internet (originally called "DARPA-Net"), and sophisticated unmanned drones. Yet it isn't just telecommunications and aerospace that DARPA has its hands in. It's nearly every sophisticated technology imaginable, including biotech. And thanks to its close relationship with the NIH, DARPA is helping to develop – and benefiting from – joint projects with this and other civilian American agencies.

In 2011, the NIH–DARPA relationship began in earnest over technology aimed at predicting drug safety for new medicines hitting the market. The NIH agreed to fund the project to the tune of $70 million over five years, while DARPA would match the NIH's funding for the program during that time.[1] But their relationship goes much deeper than this.[2] Because the NIH shares such deep ties with America's premier agency for developing covert military programs, and since the NIH is clearly so much less secure than DARPA, there is a possibility that Chinese intelligence is targeting the NIH to gain access not only to proprietary medical research done by the NIH and its fellow civilian organizations, but also to DARPA's highly secretive networks.

Why else would President Joe Biden have declared the creation of a DARPA–like agency (named ARPA-H) specifically for radical medical research under the command of the NIH? It's likely because the NIH could not ensure the safety and integrity of DARPA's classified research through its joint programs and, since the NIH is committed to maintaining its open system of research, it's easier for the NIH to replicate the DARPA radical research model without relying on DARPA itself.

Calls for a DARPA-type organization within the NIH are not new. As far back as 1997, medical professionals were arguing that the NIH's dominant peer review process was insufficient to the challenges of the modern medical world. According to Robert Cook-Deegan, writing in *Issues in Science and Technology* in winter of that year, "The chief alternative to competitive peer review are formula funding methods, based on political, historical, or performance factors, and what might be called the DARPA model, in which staff experts decide how to distribute funds." It is the conclusion of Cook-Deegan that the "DARPA approach is also a way to foster innovation."[3]

It took a few decades, but the NIH has finally embraced Robert Cook-Deegan's (and others') calls to create an in-house DARPA–like entity. The new agency within the NIH will eschew the traditional NIH model of peer review in favor of a "flat organizational structure designed to be nimble and staffed by tenure-limited program managers with a high degree of autonomy to select and fund projects using a milestone-based contract approach."[4] Scientists have also asked for the DARPA model at the NIH because they believed that the peer review process was far too conservative in its risk assessments of research. Yet as you've read, the NIH nevertheless had its tentacles on far too much radical, leap-without-looking biotech R&D. Now these experiments will have even less oversight.

Regardless, it seems that the security concerns of NIH peer review have been mitigated with the creation of this new agency. There will now be less reason for DARPA and the porous NIH to have an open kimono policy when it comes to sensitive research projects, since the NIH will have its own group doing the kind of research the NIH so often relied upon DARPA to do. However, the issue of the NIH still being vulnerable to foreign infiltration and influence remains. The Thousand Talents Program preyed upon the open-door research policy and lax security standards that the NIH employed for years. The ARPA-H program may ameliorate some of this, but it does little to address the wider threat of foreign subversion.

One avid proponent of the ARPA-H program was Dr. Terry Magnuson, who had been the head of the UNC Chapel Hill's genetics program before being fired in March 2022 for plagiarizing an NIH grant, a serious ethical violation.[5] In fact, an examination of publicly available email chains from 2017 prove that Magnuson and his colleague, Ralph S. Baric, were intimately involved in securing DARPA funding for their research (Magnuson was a leading cancer researcher whereas Baric was working on gain-of-function tests for horseshoe bats).[6] These types of eggheads are the ones now seeking to create an unaccountable, though more efficient, ARPA-H at the NIH. We need cancer research, but not unaccountable research methods.

One stark data point illustrates how unserious both the Biden Administration and the overall American scientific community are about our national biosecurity. Among Biden's first actions as President was disbanding the FBI's China Initiative.[7] Began in 2018 by President Trump, this elite counterintelligence unit was, as its name suggested, dedicated to finding and stopping

Chinese espionage directed against critical US industries, labs, government organizations, academic institutions, and individual innovators and researchers. Yet much of the scientific community here in the States loathed this group because they believed it unfairly targeted scientists of foreign (specifically, Asian or, more to the point, Chinese) backgrounds, and because the existence of this aggressive counterintelligence unit was making America less competitive than other countries.[8]

While it's true that the FBI is still taking seriously the threats of Chinese industrial espionage, it is now widening its investigation aperture to include espionage by Russia, North Korea, and Iran. This is, of course, wise. China is not the only foreign threat trying to steal critical American high-tech. But only China possesses the national leadership, will, and resources to conduct what amounts to a shadow war. By refusing to focus on the Chinese threat and dispersing its limited resources, the FBI risks missing the next big espionage hit from China.

Thousand Talents? More Like Death to America by A Thousand Paper Cuts

China's Thousand Talents Program represents the most significant danger in recent history to America's technological dominance. Beyond that, its legality is questionable. Most countries have a talent program, but those programs are almost always above-board. China's is specifically tailored to steal American intellectual property. Chinese high-tech industrial espionage costs Americans untold resources.

Clearly, during the Trump Administration, the FBI and the NIH began awakening to this threat. Under Biden's new bureaucratic ecosystem, it is not certain whether the operational tempo of the FBI (and other US security agencies) can be sustained. But for real change and real security to take shape, medical professionals and scientists will have to recognize the danger of working alongside Chinese scientists. These American researchers will have to eschew personal interest – careerist, financial, and otherwise – and commit to the greater interests of their nation's security. If the American scientific community fails to recognize the dangers, if it continues wantonly cooperating with the Thousand Talents Program, Communist China will subordinate America to its growing totalitarian empire.[9]

CHAPTER 18
KNOCKIN' ON ALL UNDER HEAVEN'S DOOR

Nanotechnology is an esoteric field, one that was traditionally relegated to the realm of science fiction. But it is now a real, serious, and growing branch of scientific inquiry. What's more, its concerns intersect with those of biotechnology. In fact, many scientists are convinced that nanotechnology "holds promise for reducing overall medical costs, improving outcomes, and adding value to drug therapy – mostly by increasing the efficiency of drug discovery, disease detection, and drug delivery."[1]

"Nanotechnology" is very small technology. The prefix "nano" should give the reader some idea as to how small: a nanometer is a billionth the length of a meter. Thus, nanotechnology deals with manipulating devices at the molecular and atomic level. First proposed as long ago as 1959, by the great physicist Dr. Richard Feynman, it was simply described as the "process in which scientists would be able to manipulate and control individual atoms and molecules."[2] In 1969, physicist Norio Taniguchi coined the term "nanotechnology." The official exploration of nanotechnology began with the 1981 creation of the scanning tunneling microscope, which allowed scientists to view individual atoms.[3] Today, scientists call the fusion of biotechnology and nanotechnology "nanobiotechnology."

Keep in mind that while the industry is still relatively new (like biotech itself), nanoparticles, incredibly small particles that are artificially produced in labs, have been used to create vaccines since the 1960s. In fact, it is from this historical use of rudimentary nanoparticles, such as liposomes, in vaccines that a large portion of the nanobiotechnology field has evolved.[4] The ability to attack a disease at the molecular level, whether it be COVID-19, the flu, or cancer, is appealing to scientists, especially in the wake of increasing antibiotic-resistant infections. Many vaccines, for example, utilize liposomes as a "versatile nanocarrier platform" to deliver the vaccine to the body. More sophisticated nanoparticles, such as lipid nanoparticles, are deployed in the current mRNA COVID-19 vaccines to prevent the breakdown of mRNA from the vaccine before it can trigger an immune response from cells.[5]

The COVID-19 vaccines represent only the beginning of nanotechnology's entrance into the growing biotechnology field. Charles Lieber, a Harvard University scientist, possibly the world's foremost expert in nanotechnology, would like you to envision a world in which nanotech stops diseases.[6] Lieber is responsible for having created the world's first "bio-compatible transistor the size of a virus," in 2011. It's part of a larger, longer-running program that he has been spearheading for decades. With countless publications and innovations under his belt, funded by numerous streams of revenue, including DARPA, Lieber comes across as his generation's Oppenheimer. But unbeknownst to almost anyone in Lieber's orbit, from 2012–2020 he was a primary player in China's ongoing Thousand Talents Program.[7]

The world would discover the truth about Lieber in 2020, when the FBI arrested and indicted him for lying to investigators about his ties to Chinese labs and the Thousand Talents Program. The Bureau also arrested two junior researchers and indicted them as part of the same probe. One, a thirty-year-old cancer researcher from China who was working at Beth Israel Labs, was arrested at Boston's Logan International Airport in possession of a vial of unknown materials hidden in his luggage. The researcher, Zaosong Zheng, claimed he stole the proprietary material to take back to his lab in China, so he could claim he invented it in his lab there, and reap the rewards of that so-called discovery. Of course, left unsaid was likely an espionage element for the benefit of China's military.

Another researcher affiliated with the DOJ's case against Lieber was a twenty-eight-year-old woman named Yanqing Ye, who had admitted to being a lieutenant in the People's Liberation Army during her interrogation by FBI officials. While working as a research assistant in the United States, PLA Lieutenant Yanqing Ye conducted multiple covert intelligence-gathering missions for her masters back in Beijing.

For almost a decade, Charlies Lieber, the Pentagon's foremost expert on nanotechnology, is believed to have opened a lab in Wuhan and overseen its operations. This lab contributed to China's own research and development of nanotechnology.[8] Along with him came a coterie of Chinese researchers who were likely military scientists sent to conduct high-tech espionage in the United States. The DOJ announced the arrests of these three individuals on the same day. They are part of a disturbing, ongoing trend.[9]

Lieber was paid handsomely for his work by China – all while receiving generous funding from the US government and his employer, Harvard University. The DOJ does not say they believe Lieber was a spy. (As an aside, if they did believe this, they probably wouldn't publicly admit it, given Lieber's standing in and access to the US defense establishment.) It does argue, however, that "corrupting levels" of money were paid to him by the Chinese government, pressuring him to do work with China, and to hide it from his American employers.

For the record, Dr. Lieber received more than $15 million in NIH grants, along with massive funding from the Pentagon, for his work. To receive that money, Dr. Lieber was required to give personal information to the government proving that he had no foreign ties or conflicts of interest. The DOJ accused Dr. Lieber of lying on those forms. According to the 2020 DOJ indictment, "Unbeknownst to Harvard University beginning in 2011, Lieber became a 'Strategic Scientist' at Wuhan University of Technology (WUT) in China and was a contractual participant in China's Thousand Talents Plan from in or about 2012 to 2017." Here's the kicker: "Under the terms of Lieber's three-year contract, WUT paid Lieber $50,000 USD per month, living expenses of up to 1,000,000 Yuan (approximately $158,000 USD at the time), and awarded him more than $1.5 million to establish a research lab at WUT."[10] For six years, Lieber earned from his illicit dealings with China $600,000 per year *plus* $158,000 in living expenses! The DOJ was right when they said that's "corrupting levels" of money.

The CCP owned America's foremost expert on nanotechnology for almost a decade – even as he was still being given access to tranches of US taxpayer funding and classified information for his work.

Lieber denies these charges. The US government repeatedly gave Lieber opportunity to come clean about his involvement with China's Thousand Talents Program. But Lieber was caught having lied multiple times to both NIH and FBI investigators. Part of Lieber's contract with China entailed him not only sharing his US taxpayer–funded proprietary research, but also identifying American Ph.D. students and other promising innovators for China to recruit into the Thousand Talents Program. Here lies the ingenious, self-perpetuating nature of the Thousand Talents Program. All Beijing had to do was to throw money at key American scientists and encourage them to recruit others. The rest took care of itself.

As for Dr. Lieber's fate, over a six-day trial, he was found guilty on two counts of having lied to federal authorities "in an attempt to hide his participation in the Chinese Thousand Talents Program." It sounds as if the Chinese had already used their "undue influence" with him, and our DOJ knew it.[11] We will never know for sure what, precisely, of the DOD's highly classified nanotechnology R&D Lieber gave to China, but we can speculate, given the scientific papers and articles he wrote during the time. Related to his breakthrough with nanotechnology from 2011 was an interest in what's known as BCI, a Brain–Computer Interface.[12] This is a technology that companies including the Elon Musk–supported firm Neural Link are working on. It is believed that this could revolutionize the way humans use technology. A working BCI could also transform the way we wage war. More disturbingly, it could give totalitarian security states, such as the PRC, a tool of literal mind control.[13]

CHAPTER 18

Nanotechnology in China

According to *Small Wars Journal*, a preeminent academic publication in the national security community, "China leads the world in nanotech startups."[14] The article goes on to elaborate how China has attracted scores of foreign multinational firms, investors, and researchers from around the world to do their nanotechnology R&D in China as opposed to the United States or the West more generally. Dr. Charles Lieber was but one of countless experts in this cutting-edge industry who chose to do business in China rather than his homeland.

In fact, there exists in the eastern Chinese city of Suzhou "the world's largest nanotech industrial zone called 'Nanopolis.'" Thanks to the CCP's heavy investment in developing advanced, concentrated infrastructure for nanotechnology in this region, "From cloning to cancer research, from sea to space exploration, [China] is using nanoscience and nanotechnology innovation to drive some of the world's biggest breakthroughs."[15] And American researchers and investors are apparently happy to help – and lie to their own government and people about it.

Let's go back to Lieber's experiments at Harvard that were documented in 2011. Back then, Lieber and his team managed to create a "bio-compatible transistor the size of a virus." The specific goal of Lieber's experiment was to create a "biological interface, in which a nanoscale device could communicate with a living organism."[16] But how can a scientist insert such a small machine into a cell without killing it? This was the brilliance of Lieber's 2011 discovery. He coated his V-shaped, hairpin nanowire with a fatty lipid layer – the same substance that that cell membranes are made of – causing the nanowire to be absorbed by the cell. Critically, this proved that manmade machines could mimic the natural processes of cells. The nanowire mimicked a natural virus inserting itself into bacteria, and it worked perfectly.

Lieber's experiment also demonstrates a potential marriage between biotechnology and nanotechnology. After all, the current mRNA COVID-19 vaccines use lipids as a tool for ensuring that the vaccine is delivered effectively to the human body. It's just another example of how the two industries feed off one another. And it explains why China is obsessed with getting access to those conducting this cutting-edge research.

In July 2019, an "international team of researchers led by Chinese scientists developed a new form of synthetic, biodegradable nanoparticle. Capable of targeting, penetrating, and altering cells by delivering the CRISPR/Cas9 gene-editing tool into a cell, the nanoparticle can be used in the treatment of some single-gene disorders, as well as other diseases including some forms of cancer."[17] Meanwhile, medical researchers at China's Nanjing University "have used nanoparticles to target and destroy abnormal proteins known to cause breast cancer."[18] Nanobiotechnology can be used for lifesaving, necessary treatments, but also for biological enhancement. At the University of Science and Technology of China, for instance, researchers gave mice "infrared night vision by injecting nanoparticles into their eyes." At the State Key Laboratory of Robotics in Shenyang, Chinese researchers used a laser to produce a tiny gas bubble that, according to Ben Halder of the popular online science magazine *Ozy*, "can be used as a tiny 'robot' to manipulate and move materials on a nanoscale with microscopic precision. The technology promises new possibilities in the field of artificial tissue creation and cloning."[19]

Halder further reports that, in 2018, Chinese researchers were responsible for 40 percent of all global scientific papers in nanotechnology. American researchers were responsible only for 15 percent. Given that nanotechnology feeds into biotechnology (as well as quantum computing, metamaterials research, semiconductor technology, and the development of artificial intelligence), the nation that dominates this field will dominate all the other key sectors of the Fourth Industrial Revolution.[20]

Beijing believes that China, with its rich, four-thousand-year-long history, must return to its dominant global position, after having been laid low in recent centuries by rapacious Western powers. The Chinese Communist Party, despite having been birthed out of Western ideology, believes it is the regime to restore China's preeminence. To accomplish this mission, China's rulers know that their country must dominate the

CHAPTER 18

global high-tech revolution. Historically, the Chinese have believed their nation represents what's known as *Tianxia*. In English, this translates to "All Under Heaven." What this means is that China's rulers have long believed that their nation was destined to be the center of the world system – that China is the greatest, most advanced nation in history and, therefore, that it is their right to rule. The CCP has taken up this cause and perpetuated it with concepts like Xi Jinping's "China Dream" of seeing China become the world's dominant superpower by the 2049 centennial of the founding of the People's Republic of China.[21] China's leaders believe they will achieve this ambitious task by dominating those scientific fields that deal with the smallest of particles: nanotechnology and biotechnology.

Short-sighted and greedy Westerners have become accomplices to a China that has imbued its scientific community with a nationalistic drive to rule the world. We're told that these American scientists are simply ignorant about China's global ambitions. People like Dr. Charles Lieber, it goes, are not spies for China, but are unknowing pawns of the regime. The Department of Justice claims that such scientists are motivated to cooperate with China only by research opportunities and financial inducements.

But it remains an open question whether Western scientists like Charles Lieber understand the full implications of what they're doing. It's bad enough if those signing on to the Thousand Talents Program are simply ignorant of the implications of their actions. But if Lieber and others today are more akin to the Soviet-sympathizing scientists of the Manhattan Project, purposely feeding sensitive information to the communists out of ideological fealty, then our problems are far more severe.

If that is indeed the truth, then US national security may be in far greater long-term danger than anyone in Washington understands. It would behoove US national security officials to review government-funded researchers' backgrounds for ties to radical, leftwing groups that might ideologically entice American researchers to share classified data with Communist China. The stakes are too high not to be more aware of this severe and growing threat.

"If you build it, they will come." Beginning in 2008, China began an ambitious and comprehensive program of expansion in advanced technology. Much of China's ongoing high-tech boom was fostered by its Thousand Talents Program and by more traditional forms of espionage targeting the West. China has struggled to match the innovative power of the United States and the West more generally. But espionage and national funding has put China in a position to compete against, and even overshadow, the United States for the rest of this century.

China's investment in its high-tech infrastructure is key. Hubs of innovation such as the aforementioned "Nanopolis" have promoted cutting-edge nanotechnology R&D. Before the "Zero COVID" policies were enacted by President Xi Jinping, Shanghai was among these dynamic innovation hubs. So, too, was Hong Kong, before Xi crushed the city beneath his jackboots. But Shenzhen, just across from Hong Kong, has been steadily rising to replace whatever lost innovation capacity Hong Kong has suffered. So, too, are Guangzhou and Guangdong, two cities quickly becoming crucial players in China's Fourth Industrial Revolution. But no city has been more important to the nanotechnology race than Wuhan.

Wuhan is home to the largest number of people with undergraduate and graduate degrees in the world. In 2012, *Foreign Policy* predicted that Wuhan would become the eleventh most dynamic city in the world by 2025.[1] In 2013, Wuhan's six major research universities spent a combined $855 million on R&D.[2]

In 1988, Wuhan's metropolitan government established the Wuhan East-Lake High-tech Zone. Colloquially known as "Optics Valley," the zone is home to intense R&D into "optoelectronics, biology, energy, and environmental conservation."[3] Within "Optics Valley" is the Wuhan National Bio-industry Base, also known as "Biolake." Here, companies focus on the development of products for biomedicine, sustainable agriculture, medical devices, bio-manufacturing, health services, and bioenergy. Established in 2008, the Biolake subsection of Optics Valley has seen prodigious growth. Incidentally, 2008 also saw the introduction of Beijing's Thousand Talents Program. It was around the same time that China began experiencing a surge in its biotech and nanotech capabilities.

Biolake has made Wuhan one of China's most competitive high-tech research and innovation hubs, helped also by generous state policies and a massive population of skilled workers. The presence of expansive, high-tech infrastructure makes cities like Wuhan appealing to foreign investors, start-ups, and talent. Of course, foreign investment in China is somewhat more complicated after COVID-19. But much of the international interest in using China as a base for high-tech R&D has yet to abate, largely because of the welcoming policy environment.

As an example, Wuhan made it a "priority to reduce administrative costs" for companies while providing "legal, financial, human resource, risk management, and intellectual property services as well as tax credits" to businesses operating in Wuhan.[4] The local government also subsidizes lab construction and office rentals for companies and innovators looking to base their operations in the city. China has become better at preventing bureaucratic red tape from stymying business ventures. By making China more attractive to these elements, its economy benefits, and its government has access to the newest innovations, while the United States and rest of the world look on with envy.

For China to achieve its goal of replacing the United States as global leader by 2049, Beijing will need to dominate the Fourth Industrial Revolution, and to do this it will need to create as dynamic of an environment as possible. What China lacks in domestic talent or investment it makes up for in creating next-generation research and innovation labs, and in enacting attractive state policy to attract foreign talent and investment. Once ensconced in places like Wuhan or Shenzhen, these companies and researchers are forever owned by the rapacious Chinese Communist juggernaut. Their research will be shared first with China, which will then use that research to enhance its own economic and military power.

Already, China's military elite are on record as discussing China's coming "biological dominance." PLA Major General He Fuchu, vice president of China's Academy of Military Sciences, claims that "Modern biotechnology and its integration with information, nano(technology), and the cognitive, etc. domains will have revolutionary influences upon weapons and equipment, the combat spaces, the forms of warfare, and military theories."[5] China's deep penetration of the US scientific community only means that the PLA's recent quest for what General He calls "biological dominance" is closer than ever before. Places like Biolake in Wuhan's Optics Valley should give American national security practitioners pause. It is the biological domain where China could inflict – and has already inflicted – immense damage on our national security.

China's Commission for Science, Technology, and Industry for National Defense (COSTIND) along with the PLA's Military Intelligence Department have an outsized influence over many of the research facilities in Wuhan and throughout China. According to one article on China's bioweapons capability, these "Chinese facilities have frequent and systematic interactions with American scientists, often aiming to absorb – ostensibly academically – advanced know-how from [foreign] scientists."[6] In other words, China behaves very much like the Borg from *Star Trek: The Next Generation*: it will absorb any distinctive person or technology from the West into its growing collective.[7] By taking the best ideas and elements of other countries – even rival states, like America – and incorporating them into its vast and growing innovation ecosystem, the CCP is becoming a leader of the Fourth Industrial Revolution and a severe challenger to the United States.

There's an organ shortage across the globe. This has been a crisis for many years, and it isn't getting any better.[1] Scientists the world over desire new, innovative ways to deal with this significant medical issue. One possible solution that's been kicking around has been to clone human organs.[2] It's one of those outside-the-box ideas that was previously relegated to the fringes of acceptable science. But thanks to the advent of CRISPR-Cas-9 and the changes in attitudes about stem cell research, as well as technological advances in the medical sciences field, this previously outlandish concept may be on the verge of becoming reality.[3]

But how to accomplish such a fanciful task?

One solution is to use stem cells to create cloned organs within pigs. The founders of Renovate Biosciences, Inc. (RBI), Bhanu Telugu and Chi-Hun Park of the University of Maryland, have perfected this technique.[4] In 2018, Bhanu Telugu pitched his idea of using stem cells to clone organs within pigs to his university department. They jumped at the chance to test his hypothesis, and that year RBI won the University of Maryland's Inventor Pitch Award. Their experiment worked as a proof-of-concept. According to the researchers, the cloned organ market could be at least "a $3 billion a year market." In the acknowledgments section of their paper elaborating how they successfully cloned organs in a pig using stem cells, Telugu and Chi-Hun Park thank the NIH, the Guangdong Provincial Key Laboratory of Genome Read and Write, the Shenzhen Engineering Laboratory for Innovative Molecular Diagnostics, the Shenzhen Peacock Plan, and the Guangdong Provincial Academician Workstation of BGI Synthetic Genomics.

Yet again, American scientists and biotech entrepreneurs are sharing their work – and receiving additional funding from Chinese entities that, as you've read, are directly tied to China's national military program. Yes, Chinese scientists contribute to these projects, but through their contributions they are given greater access to American R&D, innovators, and institutions. Rarely is this reciprocated by China toward the Americans. Recall the 2015 assessment of biodefense expert Dany Shoham: many Chinese biotech R&D facilities that have outward-facing civilian research purposes are actually covers for China's ongoing, covert biological weapons program. Such facilities seek to absorb "advanced know-how" from American labs, innovators, and biotech firms, whose research is then, unbeknownst to those American scientists, folded into China's larger, covert, ongoing bioweapons program.[5]

The last acknowledgment in the Telugu–Park paper should pique your interest. You'll remember earlier in this book I identified BGI – Beijing Genomics Institute – as possessing the world's largest genomic database. The company itself is run by a gonzo scientist who fully believes his firm will have perfected the science of cloning in the next couple of decades, to the point that it will actually clone humans. BGI is already cloning pets and animals.[6] There are programs underway in China for creating augmented cloned police dogs.[7] These cloned dogs, created by the Beijing-based Sinogene, are designed to "increase efficiency" for police K9 units and to decrease the time it takes to train such animals.[8] In fact, Sinogene is the world leader in cloning animals. The more this technology is perfected, the more profit there is to be made. It is only a matter of time before China starts cloning humans, either entirely or merely for their organs. These developments will obviously give rise to pressing ethical questions, but China has a history of ignoring just those concerns.[9]

China has gone from cloning dogs to pigs to monkeys. The successful cloning of monkeys, genetic cousins to humans, has many people worried. While it's true that there are still many hurdles to overcome between cloning monkeys and cloning human beings, we've undoubtedly moved closer. Almost everyone in the scientific community is aware of this slow but steady progression. Regarding the cloning of humans, it seems to be a question of "when," not "if" – and that's a frightening prospect.[10]

Picture, if you will, an embryonic human fetus floating in a pinkish gelatinous sack. The baby is growing, as all babies do in the womb. Yet this is not an ordinary womb. The womb exists in a lab. And its caretaker is not its mother but instead an artificial intelligence–driven robot. The scene looks like something from the original *Matrix* film.

But I'm not describing a science fiction fantasy dystopia set in the distant future, in which evil robot overlords engineer humans to harvest their bioenergy. I am describing a conceivable, relatively near-term future set by scientists at the Suzhou Institute of Biomedical Engineering and Technology. According to these scientists, who have thus far limited their research to mouse embryos, "Above the embryos [growing in the artificial wombs] is an optical device capable of magnifying the embryos and monitoring them with impressive detail, which provides key growth information to the [AI caretaker]. Based on this information, *the AI can even rank the embryos on overall health and potential, should researchers wish it* [emphasis added]."[11]

We're through the biological looking glass, Alice. And the CCP, with its total lack of morals and ethics, is perfecting the technologies needed to create Frankenstein's monster. To what end do you think China's government will use this technology, perfected with the help of unwitting American scientists and biotech firms? Will China settle for the altruistic goal of creating a surplus of human organs to save the lives of those afflicted with terminal illnesses? Even if that is the only intention, imagine the ethical can of worms it will open! Or, more dangerously, will the technology be used to create the perfect communist? To craft the elite super-soldier, perhaps, that would prove a decisive asset in any potential conflict with the United States?

You're snickering reading this. I know you are, because that was the reaction that I initially had when I started researching this as a geotechnology analyst in the national security policy field back in 2018. But the technology is developing at such a rate that we may see cloned people and augmented Chinese soldiers within our lifetime. We may soon be living in a world where our political masters select who is born and who is aborted. They may even have the capability to create life in a lab entirely on their own, and to genetically manipulate that life to do their bidding.

As was noted at the start of this work, China has already been doing limited experiments with gene-editing technology to augment selected soldiers. While he was in office, the Trump Administration's Director of National Intelligence, John Ratcliffe, wrote a *Wall Street Journal* op-ed warning of China's military use of genetic engineering.[12] Ratcliffe was so disturbed by this weaponization that he directed more DNI resources towards tracking China's unethical development of biotech.[401]

In 2015, the NIH imposed a strict moratorium on all testing in the United States involving the creation of chimeric animals.[14] But just as with the previous NIH ban on the creation of chimeric pathogens through gain-of-function tests, the American science community couldn't leave well-enough alone. They paid lip-service, just as they did with gain-of-function tests, to upholding the moratorium. Yet the promise of pig-man could not be ignored. So the American science community did what it always does: it skirted the rules. Sure, the NIH stopped funding animal chimera projects here in the States. But really, they off-loaded these riskiest components of the research to China. This at least partly explains how the RGI–UMD pig cloning experiment was able to go forward: they received support and assistance from Chinese labs and biotech firms such as BGI to complete their project.

Not all scientists were sanguine about these experiments. Many bioethicists rightly expressed concerns for such projects. But others are looking only to move the experiments ever forward. Some scientists believe the next step is to use cows and other larger animals as host, since their size would allow for human cloned organs to grow to a size that could make those organs transplantable to human beings. There is also growing interest from the scientific research community in creating human-animal hybrids with human-like intelligence for neurological studies.[15]

Yet these chimeras could attain a level of sentience that would cause an entirely new set of ethical and legal dilemmas that most scientists – to say nothing of political leaders and legal experts – are unable or unwilling adequately to address. As one leading bioethicist has cautioned, "Chimeric research [will] move [society] into areas of unforeseen consequences, for which we are totally unprepared."[16] But the scientific community has a stock answer at the ready. In the words of Stanford's leading bioethicist, Hank Greely, "In American bioethics, 'cures' is the ultimate trump card.... You play that angle [such as by saying chimeras will provide transplantable organs to dying patients] and, politically, you almost always win."[17] American biotech innovators and investors have already played that trump card to great effect.

And when they couldn't get the NIH to support their endeavors after the moratorium, they went to China. But the research these American biotech leaders were sharing with China would not be contained exclusively to the realm of civilian biotech R&D. As you've seen, the Chinese bioweapons program is covert; Beijing denies it even has such a program. But almost every leading biowarfare expert in the West knows that China's biotech sector is undergirded – and indeed led – by the People's Liberation Army. While one may note that DARPA and other elements of the US military are involved in American biotech, it is essential to understand that these military organizations are a component but not the foundation of America's biotech sector. Yes, US military labs and scientists augment specific programs and share talent with the civilian research sector. But they are not the driver of developments – or even the primary beneficiary of discoveries – as the PLA is in China's biotech space.

CHAPTER 21
BIONATIONALISM VS. BIOGLOBALISM

Biotechnology is the battlespace of the future. By all accounts, China is leading the world in biosciences. China has leapfrogged the Americans in this critical domain, thanks to its wide-ranging espionage programs. Now that China has the world's leading R&D innovation clusters, Western talent and capital are excited to utilize these world-class facilities and get in on the ground floor of China's biotech miracle.

Of course, that miracle is more akin to a curse.

COVID-19 should have been the warning to the West about the dangers of partnering with China. Sadly, the pandemic has not stopped the American scientific community from looking to Beijing as a partner in scientific development. If anything, the pandemic has encouraged Western science to ally itself more closely with China, because China is not held back by ethical prohibitions or shortfalls in funding.

China keeps benefiting from the ignorance, greed, and arrogance of Western leaders and scientists. American elites seem to believe that because the United States has been the dominant power over the course of their lives, America will always be dominant, no matter what. More dangerously, these elites assume that the rest of the world shares our values, and that all other nations want to be just like the United States. So, American elites cannot fathom the existence of a country, like China, that intends to harm the US. America's elites either don't believe or don't care that their desire to make gobs of money could diminish the United States on the world stage and install the CCP as the new global master.

What the world endured during the pandemic because of China's biotech program, and what it might yet experience for the same reason, is frightening. Steps must be taken to better protect America's intellectual property.

China's scientists ascribe to an ideology that we might call "bionationalism" – the elevation of national interests through bioscientific development. Whatever contributions Americans think China's scientists are making to a global, humanitarian cause, their true ends are for the betterment of the People's Republic of China, not least militarily.

In China's western Xinjiang province, where its largest population of ethnic Turkic Muslims – known as the Uighurs – have lived for centuries, the crime of the century has unfolded. Beginning in 2012, President Xi Jinping began a severe crackdown on what his regime viewed as "ideological deviationists." In China, communism is the only religion, and the Party's head of state, in this case Mr. Xi, is its deity. Fearing Islamist terrorism would erupt from this region, that the Uighurs would become a subversive force against the CCP, Beijing began a campaign of terror, mass imprisonment, and ethnic cleansing.

Mass internment centers, what can only be viewed as modern-day concentration camps, were established by Beijing to house an estimated 1.8 million Muslim Uighurs who lived in Xinjiang.[1] Within the walls of nightmarish prison camps, female prisoners were brutally raped repeatedly by their CCP guards, and others were tortured and even murdered.[2] Outside, an Orwellian surveillance state was established to police and monitor the people of Xinjiang.[3] In Xinjiang, Beijing's rulers have tested the limits of their techno-totalitarianism. Inside the prison camps, China's cutting-edge biotech developments were put into action.

In the People's Republic of China, someone is always watching. It isn't only to manage a national crisis such as the COVID-19 pandemic that the all-powerful state peers into the lives of its citizens. The CCP does this daily. Modern technology, which so many in the West envision as a source of prosperity and liberty, is the means of centralized control. Drones, smart-phones, computers, nationwide closed-circuit television units, internet firewalls controlled by capricious communist censors – these have all become staples in ordinary Chinese life. Now that China has become a dominant participant in the global Fourth Industrial Revolution, new technologies such as artificial intelligence, cloud computing, quantum computing, and even biotech are being brought to bear in this totalitarian surveillance effort.

In Tumxuk, a city in Xinjiang, the *New York Times* came upon troubling information that Chinese researchers were using blood from Uighurs in their concentration camps – likely against the prisoners' will – to create what's known as a DNA phenotype. In DNA phenotyping, a difficult process still in its early stages, scientists analyze genes "associated with traits like ancestry, skin color, and eye color to make predictions about what the sample donor may look like, with varying degrees of certainty."[4] DNA phenotypes have been helpful for scientists trying to determine what human beings looked like long ago. But China hopes to use DNA phenotyping as a form of biomedical surveillance. DNA phenotyping for facial reconstruction purposes is yet another method of finding Uighurs and imprisoning them based on their ethnic identity.[5] Hitler, Stalin, Mao, and Saddam could have only ever dreamed of having such a technology.

Western scientists are complicit in the ongoing imprisonment and abuse of Uighurs in Xinjiang. It was reported in 2019 that European scientists conducted studies on the DNA samples that China illegally collected from unwilling imprisoned Uighurs. In fact, "at least two scientists received funding from respected institutions in Europe [and] respected international scientific journals have published [the scientists' findings] without examining the origins of the DNA used in the studies or vetting the ethical questions raised by collecting such samples in Xinjiang."[6] Don't be surprised by this. You've read how our scientific community in the West is dominated by the dual mentalities of "trust the science" and "publish-or-perish." So long as these European scientists didn't have to get their hands dirty and only had to handle the data derived from China's abhorrent actions, it's likely the European scientists simply buried their heads in the sand and urged their colleagues to focus on results.

Anything for science, I suppose. Let us not forget that Western scientists had no problem using research on Jews, Soviets, and others who had been imprisoned during the Nazi's reign of terror in Germany.[7] While the Western scientists did not condone Nazi research methods, after the Second World War ended Western science simply ignored those methods and instead focused on the end results to augment their own, more ethical research. This remains a controversial aspect of World War II scientific history.

It's time to understand that while China proudly practices bionationalism at our expense, American scientists and corporations – even some parts of the US government – continue to believe in "borderless science" – in bioglobalism. It's going to get us all killed if we're not careful. Beijing's leaders talk about waging impending wars upon the West. In 2019, for example, Xi Jinping literally declared a "People's War" against the United States.[8] In turn, a succession of US presidents increasingly caution about the grave threat the CCP poses to the world.[9] Yet American scientists blissfully act as if these tensions have no impact on their business. They couldn't be more wrong in doing so. Biotechnology has the potential to be used and abused by Beijing in a truly hellish war, the likes of which no American has ever experienced. If COVID-19 was indeed a bioweapon – whether released accidentally or not – the dangers of the bioglobalist creed have already proven to be absurdly high. No amount of research is worth the risk.

CHAPTER 22
SPECIFIC GENETIC ATTACKS

Remember BGI, the Chinese company with the world's largest genomic database? Well, it turns out they've been very keen on getting DNA from pregnant women, not only in China but from around the world. BGI sells one of the world's most popular pre-natal tests. The product is sold in fifty-two different countries – thankfully, excluding the United States. The test, known as the "Non-Invasive Fetal TrisomY" (NIFTY), according to *Reuters*, is used by expectant mothers in their tenth week of pregnancy to "detect abnormalities such as Down syndrome in the fetus."[1] The NIFTY tests also "capture genetic information about the mother, as well as personal details such as her country, height and weight, but not her name." A staggering eight million women globally have taken the BGI NIFTY test.[2]

Years after the tests were introduced globally, it was discovered that, contrary to what BGI promised its clients, genetic data was not discarded after testing was completed.[3] The DNA of the pregnant women – and that of their unborn children – was incorporated into BGI's massive genomic database and used for extensive R&D in their growing bioinformatics program. BGI says that the only genomics data it uses in its experiments are from Chinese citizens, but this has been challenged by a coterie of researchers and reporters, including from *Reuters*. As for the Chinese women's DNA, BGI has admitted to working with the PLA to "use a military supercomputer to re-analyze their NIFTY data and map the prevalence of viruses in Chinese women, look for indicators of mental illness in them, and single out Uighur and Tibetan minorities to find links between their genes and their characteristics."[4]

It has since been confirmed, despite the public assurances of BGI, that the genetic information of women from outside of China was stored in China's massive genebank.[5] Since 2015, China's regime has refused to allow for foreign companies and researchers to access the genetic data of Chinese citizens, citing national security concerns. Yet, idiotically, the United States allows for foreign researchers – including those from China – to gain access to the genetic information of American citizens. This is such a concern that in 2021, the Senate passed the United States Innovation and Competition Act (S. 1260) which had a provision specifically aimed at preventing Chinese genomic firms like BGI from gaining access to the genomic data of American citizens.[6] While, thankfully, NIFTY is not used in the United States, we do use plenty of other technology from China, and unless Washington is far more cautious than it has been, severe consequences could arise, especially in the realm of advanced technology.

The news about BGI using a PLA military supercomputer to collate information in their massive genomic database should trouble national security experts.[7] Google's parent company, Alphabet, expressed deep concern to the US government about China's significant advances in biotechnology and artificial intelligence.[8] Google shamelessly partnered with China in developing a significant artificial intelligence infrastructure for the nation – that is, until 2019, when the company decided to forego any further involvement.[9]

Google's did not reverse its stance on AI development in China in 2019 out of patriotic duty. Google's leadership did it because Congress, along with the Trump Administration, was coming down hard on the company for refusing to work on cloud computing with the Department of Defense, while still working with the CCP on artificial intelligence. Nevertheless, Eric Schmidt was right to raise a warning flag about the threat posed by BGI's partnership with the PLA.

Your biological data is being targeted by the Chinese government for the creation of next-generation bioweapons. It is only a matter of time before Beijing gets what it wants, if it hasn't already. In 2013, BGI purchased an American genomics company and, through that purchase, gained access to contracts and partnerships with various American health institutions, according to US counterintelligence official Edward You, who spoke with the *New York Times* on this matter.[10] Unfortunately, the US government and its scientific community have been far too slow to respond. In February 2021, the National Counterintelligence and Security Center issued a warning that opens with a jarring question: Would you want your DNA or other healthcare data going to an authoritarian regime with a record of exploiting that data for repression and surveillance?[11] Nevertheless, US biotech companies, innovators, and labs – even, as you've read, US government entities and individuals – still seek to cooperate with China.

BGI entered the market promising to "industrialize" genomic research, to make biotech truly universal. The company is making good on that promise. Recall that in 2011 BGI initiated a partnership with the Children's Hospital of Philadelphia to combat childhood cancer. This, of course, could have been an easy way for BGI to gain access to the genomic data of American children afflicted with cancer, as well as to any other sensitive information the Children's Hospital of Philadelphia might have possessed.

As it turns out, the Children's Hospital of Philadelphia wasn't the only major American biotech entity to get involved with BGI. In 2013, BGI completed its controversial purchase of an American biotech start-up, Complete Genomics, Inc., based in Mountain View, California, that had created a faster and more efficient way of gene mapping. BGI wanted to purchase the company both for its intellectual property and for its exclusive client list, which included the US National Cancer Institute, an affiliate of the NIH. In 2012, when BGI expressed interest in purchasing the American genomic firm, the Treasury Department's Committee for Investment of the United States (CFIUS) upheld the initial purchase for review. CFIUS acts as a check against foreign subversion of industries critical to US national security.

But the CFIUS misses more than it catches. BGI's purchase of Complete Genomics, Inc. went through.[12] BGI thus gained access to Complete Genomics' exclusive client list, which also included Pfizer, Eli Lilly, and the Mayo Clinic, as well as its revolutionary gene mapping tool – all of which was undoubtedly absorbed into China's robust indigenous biotech industry. Today, BGI claims it can make an individual's entire genome for less than $100.

Your DNA is nothing more than data. Data that can be exploited. It can be used not only to identify you, but also to manipulate and attack you. The combination of biotech with artificial intelligence and other advanced forms of computing would allow China's military and intelligence apparatus to develop comprehensive digital profiles of individuals and nations. Theoretically, specific weaknesses within your genes could be accessed, and weapons created specifically to harm you or those like you. Think of it as precision biowarfare.

Several years ago, *The Atlantic* published a piece about how terrorists could soon have access to gene-editing technology that would allow for them to craft a specific genetic attack against the American president and assassinate him with biotechnology.[13] If security experts were afraid of al Qaeda or ISIS taking control of this technology, just think about what it might do in the hands of China.

China's proof-of-concept on this matter will likely arise from the ongoing quest to harvest as much genetic material from populations of undesirables such as the Uighurs and Tibetans. From there, it would not be surprising if, in a few years, we start witnessing the strange deaths of large numbers of Uighurs or others. What China tests on its own population will inevitably be scaled up and directed against its great American rivals.

Creating specific ethnic genetic attacks is a far more efficient method for destroying one's enemies. Just look at how poorly the Americans reacted to COVID-19, a disease that is relatively mild within the history of pandemics. We destroyed ourselves. China only had to release the plague and then watch as the world's greater superpower crumpled from within.

Now contemplate what Beijing's strategists could achieve with a far deadlier pathogen, or with a bioweapon that only kills people with traits specific to the United States. Suppose that the Chinese discovered a genetic trait common among the United States Armed Forces. Perhaps the Chinese would want to create a bioweapon to which Americans from a specific European or African or Hispanic heritage would be susceptible, sowing chaos within the US military before the Chinese launched a military attack elsewhere in the world. With the Americans distracted because of the bioweapon that China launched, the Chinese military would theoretically have free reign to do whatever it wanted. Or imagine that China's regime wanted to kill or sicken a minority group in America to gin up discord. With biotech, China could wreak havoc on an unsuspecting military long before direct battle began.

CHAPTER 23
BIOHACKED!

I want you to get used to a term: "biohacking." This is a new concept with which relatively few are familiar. It can be broadly defined as "the activity of exploiting genetic material experimentally without regard to accepted ethical standards, or for criminal purposes."[1] There are about 100,000 biohackers known globally today. In the United States, they are a subculture of people who gather yearly at places like the "Biohacker Congress," where biohackers showcase their techniques and products for enhancing their bodies, either through drugs or other means.[2] Every November, MIT holds an "International Genetically Engineered Machine (iGEM) Competition" where "teams battle one another to build the coolest synthetically altered organisms."[3]

But the name, with its derivation from "hacker," has more ominous implications. A hacker is someone who uses computers and information software to damage, steal from, or destroy an unsuspecting individual, company, or nation. Hacking is a major component of the international black market and accounts for tens of billions of dollars' worth of damages. Naturally, hacking is a drag on the economy.[4] Now, contemplate biotechnology being used in similarly illicit ways. There is the potential for total anarchy.

When you consider what the Chinese are doing with their state security and military apparatuses in the biohacking field, you realize how dangerous the current moment is. We are at a fulcrum point in the development of humankind. With the advent of nuclear weapons, most people understood that we had created the potential for mass destruction. We've yet to experience a full-blown nuclear war. But biowarfare is another matter entirely.

Given that Beijing's military conceptualizes the biotech industry as a strategic enterprise, the chances are real that a new age in bioweapons is upon us. These bioweapons could annihilate our great country. Despite whatever treaties might exist that aspire to regulate the development and use of bioweapons, the prospects of a devastating bio-war are perhaps greater than those of a nuclear war.

It isn't only destructive warfare that this dual-use biotechnology could be used for. China may yet use this newfound technology, much of which was handed over from naïve and short-sighted American scientists and corporations, for espionage. In 2015, the US government was subjected to its most egregious cyberattack in history. Officials believe China was responsible. The personal data of roughly 21.5 million Americans who had served in the US government was stolen from the Office of Personnel Management. Of this group, 5.6 million federal employees had their fingerprints stolen.[5] Many classified US government facilities and applications require the use of personalized fingerprint scanners for access. By stealing this data, China could use advanced biotech to manipulate fingerprint scanners into thinking that someone was a US federal employee when they were not.

With the biotech sector taking such prominence, and CRISPR-Cas-9 becoming so important to the world, it is only a matter of time before "biohacking" becomes ubiquitous. It's bad enough that these capabilities are being developed heedless of the ethical and moral ramifications. That the CCP is spearheading the development is scarier.

Society is so often enamored with the promises of new technology that it rarely considers the risks. Revolutionary biotech promises to save human beings from any number of diseases. Gene-doping – using CRISPR-Cas-9 to augment human beings rather than simply cure sick people – is an alluring prospect. But without proper supervision and caution, human society could suffer far more than it stands to benefit.

Again, look at the COVID-19 pandemic. If, as I suspect, the disease came from a lab because of a gain-of-function test, the pandemic should serve as a cautionary tale for the need to develop strict regulations for biotech R&D globally. Unfortunately, our laws are reactive rather than proactive. Despite the fact that many scientists can intuit the risks of developing biotech without significant oversight, they are too often blinded by the rewards.

No one wants to abandon biotech completely. It is a field that could save countless people from disease and suffering. What we need is to comprehend that gene-doping, that hacking our biology for enhancement purposes, must be tightly regulated. Meanwhile, the Biological Weapons Convention must be updated to recognize the threat that dual-use biotech, such as gain-of-function testing, poses the world. And world governments must move to stop risky biotech R&D, whether in the United States or the People's Republic of China.

This book has detailed numerous examples of scientific excess in the West, and in America specifically. You have read how scientists ignore or skirt laws and regulations to achieve their research goals, even going so far as to cooperate with Chinese labs and firms that are covers for an extensive, growing, and threatening bioweapons program. This activity must stop. Ethical standards in biotech are weak. Half the battle will be imposing a stronger mechanism of self-regulation, particularly among scientists and medical doctors. Institutions such as peer-review and internal review boards have thus far proven insufficient to the task at hand.

Outside the scientific community, Washington must discourage biotech investors from even looking at China as a potential partner. In the wake of COVID-19, as more people began questioning the suspicious origins of the coronavirus, American elites invested in maintaining the status quo argued tirelessly that it would be disastrous to untether American and Chinese research.[6] They employed abject fearmongering.[7] These arguments must be ignored at all costs.

Washington must immediately cut itself off from Chinese biotech R&D. American leaders should define any technology transfer from US firms to China as a bribe under the Foreign Corrupt Practices Act. While this may not stop all technology transfers to China, it will stop many – and it will make such moves by American biotech entrepreneurs less attractive. Meanwhile, CFIUS must categorically prevent all Chinese biotech entities – and their subsidiaries – from purchasing American biotech properties. China managed to shock US intelligence with its pervasive high-tech espionage efforts such as the Thousand Talents Program. Scientists and biotech entrepreneurs alike must be discouraged from doing any business with China. Only the federal government can do this.

America must then forge an international consensus on regulations constraining biotech research and development. Standards must be normalized. What's more, the conversation cannot be left exclusively to scientists and those in the biotech community. They will only do whatever they can to ensure that new biotech is created. The wider public and America's elected leaders must become more intricately involved in the discussion.

While the US government moves to stunt China's access to our biotech sector, Washington must also proactively encourage responsible biotech development here in the United States. A massive surge in federal investment into biotech R&D and infrastructure must be a priority. It is not simply enough for Washington to stop the flow of information and capabilities from the United States to China. Washington must encourage positive development of biotech here. To repeat, we need biotechnology. But it is essential that development of this society-altering technology be responsible and sustainable.

Lastly, the United States should seek recompense from Beijing for damages as a result of their malign role in the cover-up, and possibly creation of, COVID-19. COVID-19 constituted a Biological 9/11. After the 9/11 attacks, the family of the victims of that terrorist attack gathered together and sued the Kingdom of Saudi Arabia for damages, since so many of the hijackers who attacked us on 9/11 were Saudis. After twenty years, the families have enjoyed some headway. The point is, though, that the US federal government should sue on behalf of the more than one million Americans who have died from exposure to COVID-19. China must be made to feel the pain for having potentially created, released, and covered up this powerful disease.

China has made clear that it intends to wage a bio-war upon the United States. Now is not the time to make deals. The United States must become far more competitive domestically in the biotech R&D sector while at the same time punishing China for its perfidy. And we must develop active defenses against the coming onslaught of offensive biohacking that Beijing means to deploy against the United States. Biohacking represents the gravest threat posed to the American people. That so few are talking about it publicly, that even fewer understand its implications for society, even while China races to dominate the building blocks of all life and to manipulate life as it sees fit, is depressing.

Putting to the side the potential for deliberate weaponization, headstrong biotech development is inherently risky. Whether scientists are creating chimeric animals, conducting gain-of-function tests, or gene-editing undesirable elements from populations (what China euphemistically calls "population quality" control), the experiments could backfire to a degree we still don't fully understand.[8] Every time someone makes a genetic change to human DNA or to the makeup of illnesses, it causes unintended knock-on changes throughout the DNA. Manipulating a soldier's DNA to make him faster, stronger, or smarter could also deactivate essential components of his DNA, or it could activate some dormant gene that could retard his development and send him down an evolutionary dead-end. We are dealing with things that could fundamentally alter human life. And we are doing it in a careless and slipshod manner.

I hope that this book has opened your eyes to the dangers of continuing to develop biotech as we have done. This will require a bipartisan, concerted effort on the part of American policymakers, and it cannot be left up only to scientists. For too long the world has looked the other way as China has marched steadily to war with the West. America and her allies cannot ignore this threat any longer. American leaders must recognize the threat of biohacking and implement the program that I have described. Gene-editing technologies, absent strong regulation or firm standards of development, will destroy us. If immediate action is not taken to reign this industry in, it will destroy us.

As this is my third book, one would think that I'd have gotten good at acknowledging those who have either directly influenced this work or indirectly inspired and supported me. But I have been blessed to have so many people in my life who have supported and championed my work that there are simply too many to count. At the risk of not being able to list everyone, I would like to raise a glass to some notables, such as F. H. Buckley, a friend and mentor in political philosophy who has been extremely generous in offering his time and advice. Without his help in putting me in touch with publishers and editors, I'd have never gotten published. I'd also like to thank Gordon G. Chang and his lovely wife, Lydia, for their continued friendship and support of my work. It is a high honor to know them and to call them colleagues.

My wife, Ashley, was instrumental throughout this work and is a continued inspiration to me. Her background in medicine and genetics was an essential guide for me on this project. She was my lodestar in this ambitious project, and I appreciate all the assistance she rendered to me.

Next, I'd like the acknowledge the editors of the *New English Review*. In early 2020, I had come by inside information indicating that not all was as it seemed in regard to COVID-19. I wrote a long-form essay breaking down what I had discovered and what I knew. But no one would publish my work on this matter. In fact, with the exceptions of Bill Gertz, Gordon G.

Chang, and Steven W. Mosher, I was one of the few public figures talking about the truth behind COVID-19. I took the essay to *NER* and they happily published it (that essay was known as "China's Growing Biotech Threat"). That essay formed the foundation of what would become a three-year research project that culminated in this work. It would never have happened, though, without that initial publication.

Julie Ponzi, Ben Boychuk, and Chris Buskirk of *American Greatness* have consistently supported me in my writing endeavors. Beginning in 2018, I began warning about risky biotech R&D being conducted in China. These articles written from 2018–20 have also contributed significantly to the research of this work. I can safely say that I was the only one in any major media publication warning about the risks that untrammeled biotech R&D in China posed the world *two years before* COVID-19 erupted. The folks at *American Greatness* provided me the platform to get this message out.

I'd also like to express my deepest condolences for Carol Herman, the longtime *Washington Times* opinion page editor. Since she first began publishing my columns in the *Washington Times* in 2020, Carol was a champion of my work. Through her consistent support of my writing I was noticed by a certain group of intelligence analysts in the US military who subsequently asked me to visit their team and discuss my research on COVID-19's origins with them. Carol was a remarkable woman, and while we only interacted via email, she will be missed. Kelly Sadler, her replacement at the *Washington Times*, has been as kind and gracious in supporting and publishing my work as was Carol. For that, I am eternally grateful.

David P. Goldman (a.k.a "Spengler") has been both an inspiration and a sounding board for my ideas over the years. Recently, he invited me to join the *Asia Times*, one of the best publications for news and analysis on Asia, where I have served as a contributor for two years. His insights into financial policy as well as Chinese capabilities and intentions have been inspiring. Even on the rare occasion where we disagree, I find David to be a scholar and a gentleman.

Steve Mosher and Bill Gertz are two other individuals whose

own research into COVID-19 aided me in this effort. I now give them thanks for their ceaseless efforts to uncover the truth about the most devastating plague in my lifetime. There were others who have supported my work for years, especially my research into biotechnology. For those of you who are not listed in this limited space, please know that I owe you much. My greatest hope is that this book starts a conversation and moves our leaders, Republican and Democrat alike, to stand tougher against China and to make Beijing pay for all the Americans who have died as a result of China's perfidy.

"1987: First Human Genetic Map," *National Human Genome Research Institute.* accessed on June 15, 2022. https://www.genome.gov/25520325/online-education-kit-1987-first-human-genetic-map

"A Brief History of CRISPR-CAS-9 Genome-Editing Tools." *BiteSize Bio.* June 30, 2020. https://bitesizebio.com/47927/history-crispr/

"About the Committee." *U.S. Senate Select Committee on Intelligence.* accessed on June 22, 2022. https://www.intelligence.senate.gov/about

"About Tom." *Tom Cotton Senator for Arkansas.* accessed on June 22, 2022. https://www.cotton.senate.gov/about

Ackerman, Todd. "MD Anderson Ousts 3 Scientists Over Concerns About Chinese Conflicts of Interest." *The Houston Chronicle.* April 20, 2020.

Adams, Ben. "Chinese Scientists Become First to Test CRISPR in Humans, as 'Sputnik 2.0' Begins." *Fierce Biotech.* November 16, 2016. https://www.fierce-biotech.com/biotech/chinese-scientists-become-first-to-test-crispr-humans-as-sputnik-2-0-begins

"Adherence to and Compliance with Arms Control, Nonproliferation, Disarmament Agreements, and Commitments." *U.S. Department of State.* August 2019. p. 45. https://www.state.gov/wp-content/uploads/2019/08/Compliance-Report-2019-August-19-Unclassified-Final.pdf

"Advanced Research Projects Agency for Health (ARPA-H): Congressional Action and Selected Policy Issues." *Congressional Research Service.* April 15, 2022. p. 2. https://sgp.fas.org/crs/misc/R47074.pdf

Akst, Jeff. "Lab-Made Coronavirus Triggers Debate." *The Scientist.* November 16, 2015. https://www.the-scientist.com/news-opinion/lab-made-coronavirus-triggers-debate-34502

Allison, Graham. "China's Ready for War – Against the U.S. If Necessary." *Baltimore Sun.* August 8, 2017. https://www.baltimoresun.com/la-oe-allison-china-war-20170808-story.html

Al Sarkhi, Awaad K. "The Link Between Electrical Properties of COVID-19 and

BIBLIOGRAPHY

Electromagnetic Radiation." *Biotechnology to Combat COVID-19*. March 14, 2021. https://www.intechopen.com/chapters/75714

Andrews, Nick, Stowe, Julia, Kirsebom, Freja, et al. "Covid-19 Vaccine Effectiveness Against the Omicron (B.1.1.529) Variant," *New England Journal of Medicine*, 386: 1532–1546, April 21, 2022. https://www.nejm.org/doi/full/10.1056/NEJMoa2119451

Andrzejewski, Adam. "Rand Paul is Doing the Right Thing by Asking for Transparency in the NIH: Opinion," *Courier-Journal*, June 20, 2022. https://www.courier-journal.com/story/opinion/2022/06/20/rand-pauls-questioning-dr-faucis-nih-royalty-payments-right/7622469001/

Andrzejewski, Adam. "Substack Investigation: Fauci's Royalties and the $350 Million Royalty Payment Stream HIDDEN by NIH." *Open The Books*. May 16, 2022. https://www.openthebooks.com/substack-investigation-faucis-royalties-and-the-350-million-royalty-payment-stream-hidden-by-nih/

Ao, Zhujun, Wang, Lijun, Mendoza, J., Cheng, Keding, et al. "Incorporation of Ebola Glycoprotein into HIV Targeting Particles Facilitates Dendritic Cell and Macrophage Targeting and Enhances HIV-Specific Immune Responses." *PLOS ONE*. May 17, 2019. https://journals.plos.org/plosone/article?id=10.1371/journal.pone.0216949

Armstrong, Annalee. "Big Pharma Partnerships, Record $22.7B Investment Raise Profile of Regenerative Medicine in 2021." *Fierce Biotech*. April 5, 2022. https://www.fiercebiotech.com/biotech/big-pharma-partnerships-record-investment-raise-profile-regenerative-medicine-2021

Asia Times, The. "Webinar: You Will Be Assimilated – China's Plan to Sino-Form the World." *YouTube*. September 30, 2020. https://www.youtube.com/watch?v=IsZTqNCS7jA

Bay Area, KPIX CBS SF. "Speaker Pelosi Visits SF's Chinatown to Show Support Amid Coronavirus Fears." *YouTube*. February 24, 2020. https://www.youtube.com/watch?v=eFCzoXhNM6c

Bay Area Staff, NBC. "Nancy Pelosi Visits San Francisco's Chinatown Amid Coronavirus Concerns." *NBC*. February 24, 2020. https://www.nbcbayarea.com/news/local/nancy-pelosi-visits-san-franciscos-chinatown/2240247/

Beard, T. Randolph, Kaserman, David L., and Osterkamp, Rigmar. *The Global Organ Shortage: Economic Causes, Human Consequences, Policy Responses* (Stanford: Stanford University Press, 2013). pp. 92–112.

Becker, Jasper. "China's Mutant Monkeys: These are Just Two of the Countless Animals Used in Secret Genetic Engineering Tests in Labs – Many with Appalling Biosecurity. No Wonder So Many Experts Say Covid DID Leak from Wuhan Research Centre, Writes Jasper Becker," *Daily Mail*, June 5, 2021. https://www.dailymail.co.uk/news/article-9655357/JASPER-BECKER-No-wonder-experts-say-Covid-DID-leak-Wuhan-research-centre.html

Begley, Sharon. "First Human-Pig Chimeras Created, Sparking Hopes for Transplantable Organs – And Debate." *Stat*. January 26, 2017. https://www.statnews.com/2017/01/26/first-chimera-human-pig/

BIBLIOGRAPHY

"Beijing Genomics Institute (BGI) Forms Children's Disease Partnership with Children's Hospital of Philadelphia." *BioSpace.* November 18, 2011. https://www.biospace.com/article/releases/beijing-genomics-institute-bgi-forms-children-s-disease-partnership-with-children-s-hospital-of-philadelphia-/

Benchley, Nathaniel. "The $24 Swindle." *American Heritage.* December 1959. https://www.americanheritage.com/24-swindle

Berry, Christopher R., Fowler, Anthony, Glazer, Tamara, and MacMillen, Alec. "Evaluating the Effects of Shelter-in-Place Policies During the COVID-19 Pandemic." *PNAS.* March 25, 2021. https://www.pnas.org/doi/10.1073/pnas.2019706118#T1

Beyrer, Jack. "Ohio State Researcher Sentenced to Prison for Concealing Ties to Chinese Spy Program." *Washington Free Beacon.* May 14, 2021. https://freebeacon.com/national-security/ohio-state-researcher-sentenced-to-prison-for-concealing-ties-to-chinese-spy-program/

"Big Pharma Exec: COVID Shots are 'Gene Therapy'." *Cause Action.* November 15, 2021. https://cqrcengage.com/causeaction/app/document/36532017

"Biohacking." *Dictionary.* accessed on July 4, 2022. https://www.merriam-webster.com/dictionary/biohacking

"Biological Weapons Convention." *United Nations Office for Disarmament Affairs.* accessed on June 17, 2022. https://www.un.org/disarmament/biological-weapons/

"Biotech: What's Driving China's Biotechnology Revolution?" *The Motley Fool.* December 26, 2018. https://www.youtube.com/watch?v=EemvGxof4Ro&feature=emb_logo

"Bioterrorism Agents/Diseases." *Centers for Disease Control.* accessed on June 24, 2022. https://emergency.cdc.gov/agent/agentlist-category.asp

Blackwill, Robert D. and Harris, Jennifer M. *War by Other Means: Geoeconomics and Statecraft* (Cambridge: Harvard University Press, 2016). pp. 93–128.

Blanchard, Ben and Holland, Steve. "China to Return Seized U.S. Drone, Says Washington 'Hyping Up' Incident." *Reuters.* December 16, 2016. https://www.reuters.com/article/us-usa-china-drone/china-to-return-seized-u-s-drone-says-washington-hyping-up-incident-idUSKBN14526J

Blouin, Nathan. Email to Ralph S. Baric, et al. May 17, 2017. https://usrtk.org/wp-content/uploads/2021/12/UNC_Baric_12.30.21.pdf

Bond, Paul. "How Americans' Opinion of Dr. Anthony Fauci Has Changed Over the Past Year," *Newsweek.* June 2, 2021. https://www.newsweek.com/how-americans-opinion-dr-anthony-fauci-has-changed-over-past-year-1596690

Bremmer, Ian. "Why the Chinese and Russian Vaccines Haven't Been the Geopolitical Wins They Were Hoping For." *Time.* August 2, 2021. https://time.com/6086028/chinese-russian-covid-19-vaccines-geopolitics/

Browne, Ed. "Fauci Was 'Untruthful' to Congress About Wuhan Lab Research, New Documents Appear to Show." *Newsweek.* September 9, 2021. https://

BIBLIOGRAPHY

www.newsweek.com/fauci-untruthful-congress-wuhan-lab-research-documents-show-gain-function-1627351

Brown, Beatrice. "DNA Phenotyping Experiment on Uighurs Raises Ethical Questions About Informed Consent." *Bill of Health*. December 9, 2019. https://blog.petrieflom.law.harvard.edu/2019/12/09/dna-phenotyping-experiment-on-uighurs-raises-ethical-questions-about-informed-consent/

Burgess, Christopher. "China's Thousand Talents Program Harvests U.S. Technology and a Guilty Verdict." *Clearance Jobs*. December 22, 2021. https://news.clearancejobs.com/2021/12/22/chinas-thousand-talents-program-harvests-u-s-technology/

Butler, Declan. "Engineered Bat Virus Stirs Debate Over Risky Research." *Nature*. November 12, 2015. https://www.nature.com/articles/nature.2015.18787

Casadevall, Arturo and Shenik, Thomas. "The H5N1 Moratorium and Debate." *mBio*. No. 3. Vol. 5. October 9, 2012. https://journals.asm.org/doi/full/10.1128/mBio.00379-12

"Case Summary: Magnuson, Terry." *The Office of Research Integrity*. accessed on 6 July 2022. https://ori.hhs.gov/content/case-summary-magnuson-terry

Cathey, Libby and Pezenki, Sasha. "Fauci, Rand Paul Get into a Shouting Match Over Wuhan Lab Research." *ABC*. July 20, 2021. https://abcnews.go.com/Politics/fauci-rand-paul-shouting-match-wuhan-lab-research/story?id=78946568

Cats Roundtable, The. "Dr. Anthony Fauci 1-26-20." *SoundCloud*. January 25, 2020. https://soundcloud.com/john-catsimatidis/dr-anthony-fauci-1-26-20

Center for Development of Security Excellence. "Case Study: Fraud/Theft of Intellectual Property: Song Guo Zheng." *Defense Counterintelligence and Security Agency*. accessed on 30 June 2022. https://www.cdse.edu/Portals/124/Documents/casestudies/case-study-zheng.pdf

Chalfant, Morgan. "Trump's National Security Adviser Says China 'Covered Up' Coronavirus." *The Hill*. March 11, 2020. https://thehill.com/homenews/administration/487011-trumps-national-security-adviser-says-china-covered-up-coronavirus

Chauhan, Sharad S. "Covid 19: The Chinese Military and Maj Gen Chen Wei." *Indian Defence Review*. October 9, 2020. http://www.indiandefencereview.com/spotlights/covid-19-the-chinese-military-and-maj-gen-chen-wei/

Chavda, Vivek P., Gajjar, Normi, and Dave, Divyang J. "Darunavir Ethanolate: Repurposing an Anti-HIV Drug in COVID-19 Treatment." *European Journal of Medicinal Chemistry Reports*, Vol. 3, December 2021. https://www.sciencedirect.com/science/article/pii/S2772417421000133

Chang, Gordon G. *The Great U.S.-China Tech War* (New York: Encounter Books, 2020).

Chan, Alina. Twitter post. March 1, 2021. 9:41 am. https://twitter.com/Ayjchan/status/1366398314791514118

Chan, Alina and Ridley, Matt. *Viral: The Search for the Origin of COVID-19* (New York: Harper Perennial, 2021).

"Chemical and Biological Weapons Status at a Glance." *Arms Control Association*. March 2022. https://www.armscontrol.org/factsheets/cbwprolif

Cheng, Jonathan. "China is the Only Major Economy to Report Economic Growth for 2020." *Wall Street Journal*. January 18, 2021. https://www.wsj.com/articles/china-is-the-only-major-economy-to-report-economic-growth-for-2020-11610936187

Chen, Celia. "US-China Tech War: Beijing's Main Policy Lender Pledges US $62 Billion to Fund Tech Innovation," *South China Morning Post*, March 4, 2021. https://www.scmp.com/tech/policy/article/3124094/us-china-tech-war-beijings-main-policy-lender-pledges-us62-billion-fund

China, Death By. "Death By China: How America Lost Its Manufacturing Base (Official Version)." *YouTube*. April 10, 2016. 14:29–14:42. https://www.youtube.com/watch?v=mMlmjXtnIXI

"China: Minority Region Collects DNA from Millions." *Human Rights Watch*, December 13, 2017. https://www.hrw.org/news/2017/12/13/china-minority-region-collects-dna-millions#

China TV, New. "'People's Hero' Chen Wei: Military Medical Scientist Working on Vaccine." *YouTube*. September 12, 2020. https://www.youtube.com/watch?v=1sqDWioJn7w

"China's Collection of Genomic and Other Healthcare Data from America: Risks to Privacy and U.S. Economic and National Security." *The National Counterintelligence and Security Center*. February 2021. https://www.dni.gov/files/NCSC/documents/SafeguardingOurFuture/NCSC_China_Genomics_Fact_Sheet_2021revision20210203.pdf

"Chinese Consulate in Houston Ordered to Close by US." *BBC*. July 23, 2020. https://www.bbc.com/news/world-us-canada-53497193

"Chinese Inhalable COVID-19 Booster Vaccine Safe and Effective: Study." *Xinhua*. May 24, 2022. https://english.news.cn/20220524/6b767673ac8047b69dba9d7d8f4dd355/c.html

Chun, Tae-Wook, Murray, Danielle, Justement, Jesse S., et al., "Broadly Neutralizing Antibodies Suppress HIV in the Persistent Viral Reservoir." *Proc Natl Acad Sci USA*. September 9, 2014. 111(36): 13151–13156. https://www.ncbi.nlm.nih.gov/pmc/articles/PMC4246957/

Chu, C.Y. Cyrus and Lee, Ronald D. "Famine, Revolt, and the Dynastic Cycle." *Journal of Population Economics*. 7.351–378. 1994. https://link.springer.com/article/10.1007/BF00161472

Cillizza, Chris. "The Story Keeps Getting Worse for Andrew Cuomo." *CNN*. February 12, 2021. https://www.cnn.com/2021/02/12/politics/andrew-cuomo-nursing-homes-covid-19/index.html

"Cloned Police Dogs are Getting a Lot of Attention at the Expo." *Sinogene*. accessed on July 3, 2022. https://www.sinogene.org/cloned-police-dogs-are-getting-a-lot-of-attention-at-the-expo.html

BIBLIOGRAPHY

CNN. "Thiel: Google Has $50B, Doesn't Innovate." *YouTube.* July 17, 2012. https://www.youtube.com/watch?v=2Q26XIKtwXQ

Cobb, Matthew. "Sexism in Science: Did Watson and Crick Really Steal Rosalind Franklin's Data?" *The Guardian.* June 23, 2015. https://www.theguardian.com/science/2015/jun/23/sexism-in-science-did-watson-and-crick-really-steal-rosalind-franklins-data

Colburn, Nora. "How Can We Know the COVID-19 Vaccine Won't Have Long-Term Side Effects?" *The Ohio State University Wexner Medical Center,* 14 September 2021. https://wexnermedical.osu.edu/blog/covid-19-vaccine-long-term-side-effects

Cohen, Jon. "'Absolutely Remarkable': No One Who Got Moderna's Vaccine in Trial Developed Severe COVID-19," *NBC,* November 30, 2020. https://www.science.org/content/article/absolutely-remarkable-no-one-who-got-modernas-vaccine-trial-developed-severe-covid-19

Cohen, Jon. "Prophet in Purgatory." *Science.* November 17, 2021. https://www.science.org/content/article/we-ve-done-nothing-wrong-ecohealth-leader-fights-charges-his-research-helped-spark-covid-19

Collins, Francis M. "Dear Colleagues Letter." *Department of Health and Human Services.* August 20, 2018. https://www.insidehighered.com/sites/default/server_files/media/NIH%20Foreign%20Inf luence%20Letter%20to%20Grantees%2008-20-18.pdf

Committee on Science, Space, and Technology. "Scholars or Spies: Foreign Plots Targeting America's Research and Development." *115th Congress.* April 11, 2018. https://www.govinfo.gov/content/pkg/CHRG-115hhrg29781/pdf/CHRG-115hhrg29781.pdf

Contera, Sonia, Bernardino, Jorge de la Serna, and Tetley, Teresa D. "Biotechnology, Nanotechnology, and Medicine." *Emerging Topics in Life Science.* 4, 6. pp. 551–54. December 9, 2020. https://portlandpress.com/emergtoplifesci/article/4/6/551/227204/Biotechnology-nanotechnology-and-medicine

Cook-Deegan, Robert. "Does NIH Need a DARPA?" *Issues in Science and Technology.* Vol. 13. No. 2. Winter 1997. https://issues.org/national-institutes-health-nih-darpa-cook-deegan/

Copley, Gregory. *The New Total War of the 21st Century and the Trigger of the Fear Pandemic* (Alexandria: ISSA, 2020). pp. 247–49.

"Coronavirus: Residents 'Welded' Inside Their Own Homes in China." *LBC.* February 2, 2020. https://www.lbc.co.uk/news/coronavirus-residents-welded-inside-their-own-home/

Cotton, Tom. "Coronavirus and the Laboratories in Wuhan." *Wall Street Journal.* April 21, 2020. https://www.wsj.com/articles/coronavirus-and-the-laboratories-in-wuhan-11587486996?mod=opinion_lead_pos5

Cotton, Tom. "Coronavirus Lab-Leak-Theory Proponents Have Been Vindicated." *National Review.* June 8, 2021. https://www.nationalreview.com/2021/06/coronavirus-lab-leak-theory-proponents-have-been-vindicated/

Court, Andrew. "Both US Aircraft Carriers in the Pacific are Taken Out of
Action for Up to a MONTH After Sailors Get Infected with Coronavirus –
Giving China a Free Hand in the Region as the Pentagon Raises Threat
Level to Second Highest Setting." *Daily Mail.* March 27, 2020. https://www.
dailymail.co.uk/news/article-8161181/Sailors-aircraft-carriers-Pacific-
coronavius.html

"COVID-19 Deaths in Wuhan Seem Far Higher Than the Official Count." *The
Economist.* May 30, 2021. https://www.economist.com/graphic-detail/
2021/05/30/covid-19-deaths-in-wuhan-seem-far-higher-than-the-
official-count

Creitz, Charles. "China 'Uses Elite Capture to Pay Off' US Oligarchs Instead of
'Head-to-Head' Conflict.'" *Fox News.* February 16, 2022. https://www.fox-
news.com/media/china-elite-american-business-corporations-
biden-schweizer

Cyranoski, David. "Gene-Edited 'Micropigs' to Be Sold as Pets at Chinese Institute." *Nature.* 526. 18. September 29, 2015. https://www.nature.com/articles/
nature.2015.18448

Cyranoski, David. "Inside the Chinese Lab Poised to Study World's Most Dangerous Pathogens." *Nature.* February 23, 2017. https://www.nature.com/
articles/nature.2017.21487

Cyranoski, David. "Profile of a Killer: The Complex Biology Powering the Coronavirus Pandemic." *Nature.* May 4, 2020. https://www.nature.com/articles/
d41586-020-01315-7

Davis, Mike. "Big Tech Censorship of COVID Information Leads to Vaccine
Hesitancy." *Newsweek.* November 2, 2021. https://www.newsweek.com/
big-tech-censorship-covid-information-leads-vaccine-hesitancy-opinion-
1644051

"Deng Xiaoping's '24-Character Strategy,'" *Global Security.* accessed on
March 25, 2021. https://www.globalsecurity.org/military/world/china/
24-character.htm

Dilanian, Ken. "China Has Done Human Testing to Create Biologically
Enhanced Super Soldiers, Says Top U.S. Official." *NBC News.* December 3,
2020.https://www.nbcnews.com/politics/national-security/china-has-done-
human-testing-create-biologically-enhanced-super-soldiers-n1249914

Dobbs, Richard and Remes, Jaana. "Introducing The Most Dynamic City of
2025." *Foreign Policy.* August 13, 2012. https://foreignpolicy.com/2012/08/13/
introducing-the-most-dynamic-cities-of-2025/

"Do Stem Cells Hold the Key for the Future of Transplantation?" *National Kidney Foundation.* accessed on July 3, 2022. https://www.kidney.org/news/
newsroom/fs_new/stemcellskey

Dr. Bauchner, Conversation with. "Coronavirus Infections – More Than Just the
Common Cold." *JN Learning.* January 23, 2020. https://edhub.ama-assn.org/
jn-learning/audio-player/18197306

Dunhill, Jack. "This Artificial Womb and AI Nanny is the Future of Child

BIBLIOGRAPHY

Development, Chinese Scientists Claim." *IFL Science.* January 31, 2022. https://www.iflscience.com/this-artificial-womb-and-ai-nanny-is-the-future-of-child-development-claim-chinese-scientists-62437

Economics, Bloomberg. Twitter post. December 22, 2019. 8:31pm. https://twitter.com/economics/status/1208923079734439936?ref_src= twsrc%5Etfw%7Ct wcamp%5Etweetembed%7Ctwterm%5E1208923079734 439936%7Ctwgr%5E%7Ctwco n%5Es1_&ref_url=https%3A%2F%2Fwww. dailywire.com%2Fnews%2Famid-trump-trade-war-china-lowers-import-tariffs-on-over-850-products%3Futm_term%3Dutm_campaign%3Ddw_ conversions_subscriptions_performa ncemax_politicalutm_source %3Dadwordsutm_medium%3Dppchsa_acc%3D641146134 4hsa_cam%3 D16599826472hsa_grp%3Dhsa_ad%3Dhsa_src%3Dxhsa_tgt%3Dhsa_kw %3Dhsa_mt%3Dhsa_net%3Dadwordshsa_ver%3D3gclid%3DCjoKCQ jwntCVBhDdARIs AMEwAClZRw5mxCdluy7kQ95mjUhZ3qoc1UeUiF 1bB2Oj2hB9gSKT92TiTWoaAiKz EALw_wcB

Editorial Board, Post. "Federal Government Using Social-Media Giants to Censor Americans." *New York Post.* September 6, 2021. https://nypost.com/ 2021/09/06/federal-government-using-social-media-giants-to-censor-americans/

"Egghead." *Dictionary.* accessed on June 27, 2022. https://www.dictionary.com/ browse/egghead

Einstein, Albert. "Why Socialism?" *Monthly Review.* May 1949. https://monthly review.org/2009/05/01/why-socialism/

Eriksson, Klara K., Makia, Divine, Maier, Reinhard, et al. "Towards a Coronavirus-Based HIV Multigene Vaccine." *Clin Dev Immunol.* 13(2–4): 353–60. Jun-Dec 2006. https://pubmed.ncbi.nlm.nih.gov/17162377/

Evans, Zachary. "NYT Publishes Op-Ed by Chinese Professor Without Disclosing Ties to Beijing." *National Review.* July 23, 2020. https://www.national review.com/news/nyt-p publishes-op-ed-by-chinese-professor-who-mocked-trump-by-calling-aids-american-sexually-transmitted-disease-in-post-on-propaganda-outlet/

Evers, Mathias and Chui, Michael. "The Promise and Peril of the Bio Revolution," *Project Syndicate*, June 26, 2021. https://www.project-syndicate.org/ commentary/biological-innovation-promise-and-perils-by-matthias-evers-and-michael-chui-2021-01?h=bi2qIbNWmgTmtpEm2gn4nKlt51 RPReGGwz9uiEwPpkY%3d&

Fact Check, Reuters. "Fact Check-mRNA Vaccines are Distinct from Gene Therapy, Which Alters Recipient's Genes." *Reuters.* August 10, 2021. https://www.reuters.com/article/factcheck-covid-mrna-gene/ fact-check-mrna-vaccines-are-distinct-from-gene-therapy-which-alters-recipients-genes-idUSL1N2PH16N

Fannin, Rebecca. "The Rush to Deploy Robots in China Amid the Coronavirus Outbreak," *CNBC*, March 2, 2020. https://www.cnbc.com/2020/03/02/the-rush-to-deploy-robots-in-china-amid-the-coronavirus-outbreak.html

Feehan, Jack. "Is COVID-19 the Worst Pandemic?" *Mauritas.* 149: 56–58. July 2021. https://www.ncbi.nlm.nih.gov/pmc/articles/PMC7866842/

Feldswich-Drentrup, Hinnerk. "How the WHO Became China's Coronavirus Accomplice." *Foreign Policy.* April 2, 2020. https://foreignpolicy.com/2020/04/02/china-coronavirus-who-health-soft-power/

Fife, Robert. "Chinese Major-General Worked with Fired Scientist at Canada's Top Infectious Disease Lab." *The Globe and Mail.* September 16, 2021. https://www.theglobeandmail.com/politics/article-chinese-pla-general-collaborated-with-fired-scientist-at-canadas-top/

Fife, Robert, Chase, Steven, and Vanraes, Shannon. "Whereabouts of Fired Winnipeg Scientists at Centre of National-Security Investigation Still Unclear," *The Globe and Mail*, June 22, 2022. https://www.theglobeandmail.com/politics/article-fired-winnipeg-scientists-location-unclear/

Fifield, Ana. "'Wolf Warrior' Strives to Make China First with Coronavirus Vaccine." *Washington Post.* March 5, 2022. https://www.washingtonpost.com/world/asia_pacific/chinas-wolf-warrior-strives-to-be-first-with-coronavirus-vaccine/2020/03/19/d6705cba-699c-11ea-b199-3a9799c54512_story.html

Fiske, Warren. "Fact-Check: Did Fauci Say Coronavirus was 'Nothing to Worry About'?" *PolitiFact.* April 29, 2020. https://www.statesman.com/story/news/politics/elections/2020/04/29/fact-check-did-f fauci-say-coronavirus-was-nothing-to-worry-about/984113007/

Frederick, Eva. "New CRISPR-Based Map Ties Every Human Gene to Its Function." *MIT News.* June 9, 2022. https://news.mit.edu/2022/crispr-based-map-ties-every-human-gene-to-its-function-0609

Friess, Steve. "Concerns Over Chinese Genomics Bid." *Politico.* December 4, 2012. https://www.politico.com/story/2012/12/concerns-arise-in-chinese-bid-for-genomics-firm-084516

Friedman, George. *The Next 100 Years: A Forecast for the 21st Century* (Random House: New York, 2009).

Fuchs, Erica R.H. "The Role of DARPA in Seeding and Encouraging Technology Trajectories: Microelectronics, Integrated Photonics, and Moore's Law." *Carnegie Mellon University.* accessed on July 1, 2022. https://web.mit.edu/iso8/pdf/Fuchs_slides.pdf

Gambill, Jason. "China and Russia are Waging Irregular Warfare Against the United States: It Is Time for a U.S. Global Response, Led by Special Operations Command." *Joint Intermediate Force Capabilities Office.* November 15, 2021. https://jnlwp.defense.gov/Press-Room/In-The-News/Article/2857039/china-and-russia-are-waging-irregular-warfare-against-the-united-states-it-is-t/

Garde, Damien. "Biden Maps Out a 'Moonshot' Approach to Cancer with Plans to 'Break Down Silos' in R&D." *Fierce Biotech.* January 13, 2016. https://www.fiercebiotech.com/r-d/biden-maps-out-a-moonshot-approach-to-cancer-plans-to-break-down-silos-r-d

Gertz, Bill. "China's Intelligence Networks in United States Include 25,000 Spies." *Washington Free Beacon.* July 11, 2017. https://freebeacon.com/national-security/chinas-spy-network-united-states-includes-25000-intelligence-officers/

Gertz, Bill. "Coronavirus Link to China Biowarfare Program Possible, Expert Says." *Washington Times.* January 26, 2020. https://www.washingtontimes.com/news/2020/jan/26/coronavirus-link-to-china-biowarfare-program-possi/?fbclid=IwAR2SobWLJgqGWoXeTDoGjSftzTKF6bBeenCKSzW6d7wK65butbxf vCKeA3M

Gertz, Bill. *How China's Communist Party Made the World Sick* (New York: Encounter, 2020).

Gilbert, Natasha and Kozlov, Max. "The Controversial China Initiative is Ending – Researchers are Relieved." *Nature.* February 24, 2022. https://www.nature.com/articles/d41586-022-00555-z

Gillum, David and Moritz, Rebecca. "Why Gain-of-Function Research Matters." *The Conversation.* accessed on June 20, 2022. https://theconversation.com/why-gain-of-function-research-matters-162493

Glasner, Joanna. "Where Google, One of the Most Active Health Care Investors, Puts Its Capital." *Crunchbase.* December 20, 2021. https://news.crunchbase.com/startups/google-health-care-startups-investment-under-the-hood/

Goldman, David P. "China Suppressed Covid-19 with AI and Big Data." *Asia Times.* March 3, 2020. https://asiatimes.com/2020/03/china-suppressed-covid-19-with-ai-and-big-data/

Goldman, David P. "Trade Wars Part Two – The Empire Strikes Back." *Asia Times.* August 6, 2019. https://asiatimes.com/2019/08/trade-wars-part-two-the-empire-strikes-back/

Goldman, David P. *You Will Be Assimilated: China's Plan to Sino-Form the World* (New York: Post Hill Press, 2020).

Gore, D'Angelo. "Some Posts About NIH Royalties Omit Fauci Statement That He Donates His Payments." *Fact Check.* May 20, 2022. https://www.factcheck.org/2022/05/scicheck-some-posts-about-nih-royalties-omit-that-fauci-said-he-donates-his-payments/

Gormley, Brian. "Venture Investment in Crispr Gene Editing Spurs Innovation, Hunt for IP." *Wall Street Journal.* March 10, 2022. https://www.wsj.com/articles/venture-investment-in-crispr-gene-editing-spurs-innovation-hunt-for-ip-11646910000

"Grassley, Johnson Release Bank Records Tying Biden Family to CCP-Linked Individuals and Companies." *Chuck Grassley.* March 29, 2022. https://www.grassley.senate.gov/news/remarks/grassley-johnson-release-bank-records-tying-biden-family-to-ccp-linked-individuals-and-companies

Gross, Terry. "CRISPR Scientist's Biography Explores Ethics of Rewriting the Code of Life." *NPR.* March 8, 2021. https://www.npr.org/transcripts/974751834

Halder, Ben. "How China is the Future of Nanoscience." *Ozy*, February 3, 2020.

https://www.ozy.com/the-new-and-the-next/cloning-to-cancer-china-is-driving-the-future-of-small-science/256094/

Hammes, T.X. "Offshore Control: A Proposed Strategy for an Unlikely Conflict." *Strategic Forum: National Defense University.* June 2012. https://ndupress.ndu.edu/Portals/68/Documents/stratforum/SF-278.pdf

Hanania, Richard. "It Isn't Your Imagination: Twitter Treats Conservatives More Harshly Than Liberals." *Quillette.* February 12, 2019. https://quillette.com/2019/02/12/it-isnt-your-imagination-twitter-treats-conservatives-more-harshly-than-liberals/

"Harvard University Professor and Two Chinese Nationals Charged in Three Separate China Related Cases." *Department of Justice.* January 28, 2020. https://www.justice.gov/opa/pr/harvard-university-professor-and-two-chinese-nationals-charged-three-separate-china-related

Hauck, Katharina and Mehta, Roshni. "The Economic Cost of China's Lasting Zero-COVID Strategy." *Think Global Health.* April 28, 2022. https://www.thinkglobalhealth.org/article/economic-cost-chinas-lasting-zero-covid-strategy

Helmenstine, Anne Marie. "The Differences Between DNA and RNA." *ThoughtCo.* February 2, 2020. https://www.thoughtco.com/dna-versus-rna-608191

Hessel, Andrew, Goodman, Marc, and Kotler, Steven. "Hacking the President's DNA." *The Atlantic.* November 2012. https://www.theatlantic.com/magazine/archive/2012/11/hacking-the-presidents-dna/309147/

Hutzler, Alexandra. "New Report Finds Cuomo's Controversial Nursing Home Guidance 'May Have Made a Bad Situation Worse.'" *Newsweek.* February 19, 2021. https://www.newsweek.com/new-report-finds-cuomos-controversial-nursing-home-guidance-may-have-made-bad-situation-worse-1570630

Hvistendahl, Mara. "Exclusive: Major U.S. Cancer Center Ousts 'Asian' Researchers After NIH Flags Their Foreign Ties." *Science Magazine.* April 19, 2019. https://www.sciencemag.org/news/2019/04/exclusive-major-us-cancer-center-ousts-asian-researchers-after-nih-flags-their-foreign

Hvistendahl, Mara. *The Scientist & the Spy: A True Story of China, the FBI, and Industrial Espionage* (New York: Riverhead Books, 2021).

Ibbetson, Ross. "Did Coronavirus Originate in Chinese Government Laboratory? Scientists Believe Killer Disease May Have Begun in Research Facility 300 Yards from Wuhan Wet Fish Market." *Daily Mail.* February 16, 2020. https://www.dailymail.co.uk/news/article-8009669/Did-coronavirus-originate-Chinese-government-laboratory.html?fbclid=IwAR1nml3TbjmIe2jt_SCgUoprGLb_EN5YMyftq9cM-UfEwHo-3LdoZNzoum8

Ignatius, David. "How Did COVID-19 Begin? Its Initial Origin Story is Shaky." *Washington Post.* April 2, 2020. https://www.washingtonpost.com/opinions/global-opinions/how-did-covid-19-begin-its-initial-origin-story-is-shaky/2020/04/02/1475d488-7521-11ea-87da-77a8136c1a6d_story.html

IIEA. "Eamonn Fingleton – The U.S. In the Coming Era of Chinese Dominance – 29 April 2014." *YouTube.* 14:08. April 30, 2014. https://www.youtube.com/watch?v=IdN_thfIiEg

Industries, Huntington Ingalls. *Twitter.* February 27, 2015. 3:13 pm. https://twitter.com/HIIndustries/status/571402713201836032

Institute, The Aspen. "Ron Klain on Coronavirus," *YouTube.* March 20, 2020. https://www.youtube.com/watch?v=h8KZ3F7JwT0

Isenstadt, Alex. "Cotton Gathers Big Donors to Talk 2024 Presidential Race." *Politico.* June 14, 2022. https://www.politico.com/news/2022/06/14/cotton-big-donors-2024-presidential-r race-00039476

Isaacson, Walter. *The Code Breaker: Jennifer Doudna, Gene Editing, and the Future of the Human Race* (New York: Simon & Schuster, 2021).

Isaacson, Walter. "mRNA Technology Gave Us the First COVID-19 Vaccine. It Could Also Upend the Drug Industry." *Time.* January 11, 2021. https://time.com/5927342/mrna-covid-vaccine/

Italiano, Laura. "Coronavirus 'Whistleblower' Nurse Says China Has 90,000 Sick." *New York Post.* January 26, 2020. https://nypost.com/2020/01/26/coronavirus-whistleblower-nurse-says-china-has-90000-sick/

Jacobsen, Rowan. "Inside the Risky Bat-Virus Engineering That Links America to Wuhan." *MIT Technology Review.* June 29, 2021. https://www.technologyreview.com/2021/06/29/1027290/gain-of-function-risky-bat-virus-engineering-links-america-to-wuhan/

Jakobsen, Rasmus Kragh. "First Chinese CRISPR Gene Therapy Trial Demonstrates Safety." *Crispr News Medicine.* April 28, 2020. https://crisprmedicinenews.com/news/first-chinese-crispr-gene-therapy-trial-demonstrates-safety/

Jee, Charlotte. "China is Using DNA Samples to Try to Re-Create the Faces of Uighurs." *Technology Review.* December 3, 2019. https://www.technologyreview.com/2019/12/03/102429/china-is-using-dna-samples-to-try-to-recreate-the-faces-of-uighurs/

Jia, Hepeng. "What is China's Thousand Talents Plan?" *Nature Jobs Career Guide.* 2018. https://media.nature.com/original/magazine-assets/d41586-018-00538-z/d41586-018-00538-z.pdf

Kaiser, Jocelyn. "NIH Says Grantee Failed to Report Experiment in Wuhan That Created a Bat Virus That Made Mice Sicker," *Science,* October 21 2021. https://www.science.org/content/article/nih-says-grantee-failed-report-experiment-wuhan-created-bat-virus-made-mice-sicker

Kania, Elsa B. and Laskai, Lorand. "Myths and Realities of China's Civil-Military Fusion Strategy." *CNAS.* January 28, 2021. https://www.cnas.org/publications/reports/myths-and-realities-of-chinas-military-civil-fusion-strategy

Kania, Elsa B. and Vorndick, Wilson. "China's Military Biotech Frontier: CRISPR, Military-Civil Fusion, and the New Revolution in Military Affairs." *Jamestown Foundation.* Vol. 19, Issue 18. October 8, 2019. https://

jamestown.org/program/chinas-military-biotech-frontier-crispr-military-civil-fusion-and-the-new-revolution-in-military-affairs/

Kania, Elsa B. and Vorndick, Wilson. "Weaponizing Biotech: How China's Military is Preparing for a 'New Domain of Warfare.'" *Defense One.* August 14, 2019.https://www.defenseone.com/ideas/2019/08/chinas-military-pursuing-biotech/159167/

Kassam, Natasha. "Great Expectations: The Unraveling of the Australia-China Relationship." *Brookings Institute*, July 20, 2020. https://www.brookings.edu/articles/great-expectations-the-unraveling-of-the-australia-china-relationship/

Keaton, Jamey. "Biden's US Revives Support for WHO, Reversing Trump Retreat." *AP News.* January 21, 2021. https://apnews.com/article/us-who-support-006ed181e016afa55d4cea30af236227

Kelley, Alexandra. "Fauci: Why the Public Wasn't Told to Wear Masks When the Coronavirus Pandemic Began." *The Hill.* June 16, 2020. https://thehill.com/changing-america/well-being/prevention-cures/502890-fauci-why-the-public-wasnt-told-to-wear-masks/

Kharpal, Arjun. "China's Baidu is in Talks to Raise Up to $2 Billion to Launch a Stand-Alone Biotech Company." *CNBC.* September 10, 2020. https://www.cnbc.com/2020/09/10/baidu-raising-money-for-biotech-firm-that-uses-artificial-intelligence-.html

Khatib, Hadi. "Cloning Pets and Animals Now Common Practice. Is Cloning Humans Next?" *AME Info.* August 27, 2019. https://www.ameinfo.com/industry/healthcare/cloning-pets-and-animals-now-common-practice-is-cloning-humans-next

Kitchen, Klon and Drexel, Bill. "Pull US AI Research Out of China." *Defense One.* August 10, 2021. https://www.defenseone.com/ideas/2021/08/pull-us-ai-research-out-china/184359/

"Klaus Fuchs." *Atomic Heritage Foundation.* accessed on June 27, 2022. https://www.atomicheritage.org/profile/klaus-fuchs

Kortepeter, Mark. "A Defense Expert Explores Whether the COVID-19 Coronavirus Makes a Good Bioweapon." *Forbes.* August 21, 2020. https://www.forbes.com/sites/coronavirusfrontlines/2020/08/21/a-defense-expert-explores-whether-the-covid-19-coronavirus-makes-a-good-bioweapon/?sh=132545837ece

Kunzmann, Kevin. "WHO, China Report Suggests COVID-19 Passed from Bats to Humans Through Another Animal." *Contagion Live.* March 29, 2021. https://www.contagionlive.com/view/who-china-report-covid-19-passed-bats-humans-animal

Kuo, Lily. "China Becomes First Major Economy to Recover from Covid-19 Pandemic." *The Guardian.* October 19, 2020. https://www.theguardian.com/business/2020/oct/19/china-becomes-first-major-economy-to-recover-from-covid-19-pandemic

Lang, Katharine. "Another Approved Malaria Medicine Shows Potential

Against COVID-19," *Medical News Today*. December 17, 2021. https://www.medicalnewstoday.com/articles/another-approved-malaria-medicine-shows-potential-against-covid-19

Lau, Susana K.P. Woo, Patrick C.Y., Li, Kenneth S.M., and Yuen, Kwok-Yung. "Severe Acute Respiratory Syndrome Coronavirus-Like Virus in Chinese Horseshoe Bats." *PNAS*. 102 (39) 14040-14045. September 16, 2005. https://www.pnas.org/doi/10.1073/pnas.0506735102

Lee, Kai-Fu. *AI Superpowers: China, Silicon Valley, and the New World Order* (New York: Houghton Mifflin Harcourt, 2018).

Leonnig, Carol and Rucker, Philip. *I Alone Can Fix It: Donald J. Trump's Catastrophic Final Year* (Penguin Press: New York, 2021).

"Li Wenliang: 'Wuhan Whistleblower' Remembered One Year On." *BBC*. February 6, 2021. https://www.bbc.com/news/world-asia-55963896

Ling, Justin. "A Brilliant Scientist Was Mysteriously Fired from a Winnipeg Virus Lab. No One Knows Why." *MacLean's*. February 15, 2022. https://www.macleans.ca/longforms/winnipeg-virus-lab-scientist/

Liu, Angus. "China's CanSino Bio Advances COVID-19 Vaccine into Phase 2 on Preliminary Safety Data." *Fierce Pharma*. April 10, 2020. https://www.fiercepharma.com/vaccines/china-s-cansino-bio-advances-covid-19-vaccine-into-phase-2-preliminary-safety-data

Lu, You, Xue, Jianxin, Mok, Tony, et al. "Safety and Feasibility of CRISPR-Edited Cells in Patients with Refractory Non-Small-Cell Lung Cancer." *Nature Medicine*. 26. Pp. 732– 40. April 27, 2020. https://www.nature.com/articles/s41591-020-0840-5

Maçães, Bruno. "How China Could Become the World's Largest Economy Much Sooner Than Expected." *Big Think*. March 7, 2022. https://bigthink.com/the-future/geopolitics-for-the-end-time/

Macias, Amanda. "FBI Arrests Chinese Researcher for Visa Fraud After She Hid at Consulate in San Francisco." *CNBC*. July 24, 2020. https://www.cnbc.com/2020/07/24/chinese-researcher-arrested-after-she-hid-at-consulate-in-san-francisco.html

Mack, John. "Nanotechnology: What's in It for Biotech?" *Biotechnology Healthcare*. 2(6): 29-61. Pp. 35–36. December 2005. https://www.ncbi.nlm.nih.gov/pmc/articles/PMC3571017/

Magnuson, Terry. "NIH On a Path to Add a DARPA-Like Model." *UNC Research*. July 1, 2021. https://research.unc.edu/2021/07/01/nih-on-a-path-to-add-a-darpa-like-model/

Makichuk, Dave. "French Prof Sparks Furor with Lab Leak Claim." *Asia Times*. April 18, 2020. https://asiatimes.com/2020/04/french-prof-sparks-furor-with-lab-leak-theory/

Manongdo, Jennifer. "China Could Harvest US Athletes' DNA at Winter Olympics, Lawmaker Warns." *International Business Time*. June 16, 2021. https://www.ibtimes.com/china-could-harvest-us-athletes-dna-beijing-winter-olympics-lawmaker-warns-3227033

Mao, Lei, Zhang, Yu, and Huang, Leaf. "mRNA Vaccine for Cancer Immunotherapy." *Molecular Cancer.* 41. February 25, 2021. https://molecular-cancer.
biomedcentral.com/articles/10.1186/s12943-021-01335-5

Marino, Lori. "We've Created Human-Pig Chimeras – But We Haven't Weighed
the Ethics." *Stat.* January 26, 2017. https://www.statnews.com/2017/01/26/
chimera-humans-animals-ethics/

Markel, Howard. "Patents, Profits, and the American People – The Bayh-Dole
Act of 1980." *New England Journal of Medicine.* August 29, 2013. https://
www.nejm.org/doi/full/10.1056/nejmp1306553

Markets, Research and. "Global Market for Bioelectronic Medicine to 2025 –
Key Role That Government Funding Agencies Such as the NIH and
DARPA are Playing in the Development of Industry." *CISION.* June 14, 2019.
https://www.prnewswire.com/news-releases/global-market-for-bio
electronic-medicine-to-2025---key-role-that-government-funding-
agencies-such-as-the-nih-and-darpa-are-playing-in-the-development-of-
the-industry-300867735.html

Markson, Sharri. *What Really Happened in Wuhan: A Virus Like No Other, Countless Infections, Millions of Deaths* (New York: Harper Collins, 2021).

Mayer, Jane. "How Russia Helped Swing the Election for Trump." *New Yorker.*
September 24, 2018. https://www.newyorker.com/magazine/2018/10/01/
how-russia-helped-to-swing-the-election-for-trump

McFadden, Robert D. "David Greenglass, the Brother Who Doomed Ethel
Rosenberg, Dead at 92." *New York Times.* October 14, 2014. https://www.
nytimes.com/2014/10/15/us/david-greenglass-spy-who-helped-seal-the-
rosenbergs-doom-dies-at-92.html

McGray, Douglas. "Biotech's Black Market." *Mother Jones,* September/October
2002. https://www.motherjones.com/politics/2002/09/biotechs-black-
market/

McGuiness, Ross. "WHO Responds to Claims Wuhan Lab Worker Could Be
COVID Patient Zero." *Yahoo! News.* August 13, 2021. https://nz.news.yahoo.
com/who-wuhan-lab-worker-covid-patient-zero-120940428.html

Medford, Heidi. "Luc Montagnier (1932–2022)." *Nature.* March 4, 2022. https://
www.nature.com/articles/d41586-022-00653-y

Melchor, Annie. "Canadian Official Reprimanded for Withholding Winnipeg
Lab Info." *The Scientist.* June 23, 2021. https://www.the-scientist.com/
news-opinion/canadian-official-reprimanded-for-withholding-
winnipeg-lab-info-68919

Meso, Andrew Isaac. "Another Diversity Problem – Scientists' Politics." *Nature.*
December 8, 2020. https://www.nature.com/articles/d41586-020-03479-
8?WT.ec_id=NATURE-20201210&utm_source=nature_etoc&utm_medium=
email&utm_campaign=20201210&sap-outbound-id=A3144545E3EB42B-
49FA04221AE0D04B3DA89D153

Mervis, Jeffrey. "U.S. Prosecutor Leading China Probe Explains Effort to That
Led to Charges Against Harvard Chemist." *Science.* February 3, 2020.

https://www.science.org/content/article/us-prosecutor-leading-china-probe-explains-effort-led-charges-against-harvard-chemist

Mitchell, Lincoln. "Dr. Fauci and Sen. Rand Paul's Recent Tense Exchange Over Covid Sends a Message." *NBC*. January 12, 2022. https://www.nbcnews.com/think/opinion/dr-fauci-sen-rand-paul-s-recent-tense-exchange-over-ncna1287366

Mizokami, Kyle. "Here's Every Aircraft Carrier in the World." *Popular Mechanics*. December 16, 2020. https://www.popularmechanics.com/military/navy-ships/g2412/a-global-roundup-of-aircraft-carriers/

Montagnier, Luc, Del Giudice, Emilio, Aissa, Jamal, et al. "Transduction of DNA Information Through Water and Electromagnetic Waves." *Pub Med*. 2015. 34(2): 106–12. https://pubmed.ncbi.nlm.nih.gov/26098521/

Moran, Nuala. "Researchers Trace COVID-19's Family Tree to Battle Outbreak and 'Infodemic.'" *Bio World*. February 14, 2020. https://www.bioworld.com/articles/433087-article-headline

Mosher, Steven W. "China's Deception Over COVID-19's Origins Gets More Outrageous Every Day." *New York Post*. March 27, 2021. https://nypost.com/2021/03/27/chinas-deception-over-covid-origins-more-outrageous-every-day/

Mosher, Steven W. "Don't Buy China's Story: The Coronavirus May Have Leaked from a Lab." *New York Post*. February 22, 2020. https://nypost.com/2020/02/22/dont-buy-chinas-story-the-coronavirus-may-have-leaked-from-a-lab/

Mosher, Steven W. "Here's All the Proof Biden Needs to Conclude COVID-19 Was Leaked from a Lab." *New York Post*. July 24, 2021. https://nypost.com/2021/07/24/heres-all-the-proof-biden-needs-to-conclude-covid-19-was-leaked-from-a-lab/

Moutinho, Sofia. "Chinese COVID-19 Vaccine Maintains Protection in Variant-Plagued Brazil." *Science*. April 9, 2021. https://www.science.org/content/article/chinese-covid-19-vaccine-maintains-protection-variant-plagued-brazil

Nakamura, David, Leonnig, Carol D., and Nakashima, Ellen. "Matthew Pottinger Faced Communist China's Intimidation as a Reporter. Now He's at the White House Shaping Trump's Hard Line Policy Toward Beijing." *Washington Post*. April 29, 2020. https://www.washingtonpost.com/politics/matthew-pottinger-faced-communist-chinas-intimidation-as-a-reporter-hes-now-at-the-white-house-shaping-trumps-hard-line-policy-toward-beijing/2020/04/28/5fb3f6d4-856e-11ea-ae26-989cfce1c7c7_story.html

Nakazawa, Katsuji. "China Knew of Lab Safety Concerns from Last Year." *Nikkei Asia*. April 30, 2020. https://asia.nikkei.com/Editor-s-Picks/China-up-close/China-knew-of-lab-safety-concerns-from-last-year

Needham, Kirsty and Baldwin, Clare. "China's Gene Giant Harvests Data from Millions of Women." *Reuters*. July 7, 2021. https://www.reuters.com/investigates/special-report/health-china-bgi-dna/

Needham, Kirsty. "Exclusive: China Gene Firm Providing Worldwide COVID

Tests Worked with Chinese Military." *Reuters.* January 30, 2021. https://www.reuters.com/article/us-c china-genomics-military-exclusive/exclusive-china-gene-firm-providing-worldwide-covid-tests-worked-with-chinese-military-idUSKBN29Z0HA

Needham, Kirsty. "Special Report: COVID Opens New Doors for China's Gene Giant." *Reuters.* November 5, 2020. https://www.reuters.com/article/us-health-coronavirus-bgi-specialreport/special-report-covid-opens-new-doors-for-chinas-gene-giant-idUSKCN2511CE

Nesbit, Jeff. "Google's True Origin Partly Lies in CIA and NSA Research Grants for Mass Surveillance." *QZ.* December 8, 2017. https://qz.com/1145669/googles-true-origin-partly-lies-in-cia-and-nsa-research-grants-for-mass-surveillance/

Newsmax. "Dr. Anthony Fauci's Thoughts on COVID-19 from January 2020." *Twitter.* April 6, 2020. 11:12 AM. https://twitter.com/newsmax/status/1247180304823062529?lang=en

"NIH, DARPA, and FDA Collaborate to Develop Cutting-Edge Technologies to Predict Drug Safety." *National Institutes of Health.* September 16, 2011. https://www.nih.gov/news-events/news-releases/nih-darpa-fda-collaborate-develop-cutting-edge-technologies-predict-drug-safety

"NIH Research Involving Introduction of Human Pluripotent Cells into Non-Human Vertebrate Animal Pre-Gastrulation Embryos." *National Institutes of Health.* September 23, 2015. https://grants.nih.gov/grants/guide/notice-files/NOT-OD-15-158.html

Nimmo, Jamie. "Zuckerberg Fund Builds $100m Stake in Biotech Company Working with GlaxoSmithKline on Coronavirus Cures." *This is Money.* May 30, 2020. https://www.thisismoney.co.uk/money/news/article-8372251/Zuckerberg-fund-bets-100m-coronavirus-cures.html

Ning, Wang, Shi-Yue, Li, Xing-Lou, Yang, et al. "Serological Evidence of Bat SARS-Related Coronavirus Infection in Humans, China." *Virologica Sinica.* November 21, 2017. https://www.ecohealthalliance.org/wp-content/uploads/2018/03/Virologica-Sinica-SARSr.pdf

Normile, Dennis. "Chinese Scientists Who Produced Genetically Altered Babies Sentenced to 3 Years in Jail." *Science Magazine.* December 30, 2019. https://www.sciencemag.org/news/2019/12/chinese-scientist-who-produced-genetically-altered-babies-sentenced-3-years-jail

Normile, Dennis. "CRISPR Bombshell: Chinese Researcher Claims to Have Gene-Edited Twins," *Science Magazine.* November 26, 2018. https://www.sciencemag.org/news/2018/11/crispr-bombshell-chinese-researcher-claims-have-created-gene-edited-twins

Office, NYC Mayor's. *Twitter.* February 13, 2020. 6:16 pm. https://twitter.com/NYCMayorsOffice/status/1228095506368344066

Ollstein, Alice Miranda. "Trump Halts Funding to the World Health Organization." *Politico.* April 14, 2020. https://www.politico.com/news/2020/04/14/trump-world-health-organization-funding-186786

O'Neill, Natalie. "COVID-19 Lab Leak Theory a 'Probable Hypothesis,' WHO Scientist Says in Stunning Reversal." *New York Post.* August 12, 2021. https://nypost.com/2021/08/12/covid-19-lab-leak-theory-probable-hypothesis-who-scientist/

O'Neill, Natalie. "WHO Scientist Wants Closer Look at Wuhan Lab That Moved Days Before COVID-19 Outbreak." *New York Post.* August 13, 2021. https://nypost.com/2021/08/13/who-scientist-eyes-on-wuhan-lab-that-moved-before-pandemic/amp/

"Oppenheimer Security Hearing." *Atomic Heritage Foundation.* July 7, 2014. https://www.atomicheritage.org/history/oppenheimer-security-hearing

Oshin, Olafimihan. "Twitter Bans Conservative Author Alex Berenson." *The Hill.* August 29, 2021. https://thehill.com/homenews/media/569908-twitter-bans-conservative-author-Alex-Berenson/

Owen, Glen. "WHO Chief 'Believes COVID Did Leak from Wuhan Lab' After a 'Catastrophic Accident' in 2019 Despite Publicly Maintaining 'All Hypotheses Remain on the Table.'" *Daily Mail.* June 18, 2022. https://www.dailymail.co.uk/news/article-10930501/WHO-chief-believes-Covid-DID-leak-Wuhan-lab-catastrophic-accident-2019.html?fbclid=IwAR0V2YOf WlMsZycNi5aHDacEkHgG1BpiTcwPen97uyUCES5IF Bt3d52kuqU

Owens, Mackubin. "Countering China's Grand Strategy." *American Greatness.* November 28, 2021. https://amgreatness.com/2021/11/28/countering-chinas-grand-strategy/

Parker, Ashley, Dawsey, Josh, Viser, Matt, and Scherer, Michael. "How Trump's Erratic Behavior and Failure on Coronavirus Doomed His Reelection." *Washington Post.* November 7, 2020. https://www.washingtonpost.com/elections/interactive/2020/trump-pandemic-coronavirus-election/

Park, Chi-Hun, Jeoung, Young-Hee, uh, Kyung-Jung et al. "Extraembyonic Endoderm (XEN) Cells Capable of Contributing to Embryonic Chimeras Established from Pig Embryos." *Stem Cell Reports.* Vol. 16. Issue 1. December 17, 2020, pp. 212–23. https://www.cell.com/stem-cell-reports/fulltext/S2213-6711(20)30459-8?_returnURL=https%3A%2F%2Flinkinghub.elsevier.com%2Fretrieve%2Fpii%2FS221 3671120304598%3Fshowall%3Dtrue

Pauls, Karen. "'Wake-Up Call for Canada': Security Experts Say Case of 2 Fired Scientists Could Point to Espionage," *CBC,* June 10, 2021. https://www.cbc.ca/news/canada/manitoba/winnipeg-lab-security-experts-1.6059097

Pauls, Karen and Ivany, Kimberly. "Mystery Around 2 Fired Scientists Points to Larger Issues at Canada's High-Security Lab, Former Colleagues Say." *CBC.* July 8, 2021. https://www.cbc.ca/news/canada/manitoba/nml-scientists-speak-out-1.6090188

Pear, Robert. "U.S. Officials Warn Health Researchers: China May Be Trying to Steal Your Data." *New York Times.* January 6, 2019. https://www.nytimes.com/2019/01/06/us/politics/nih-china-biomedical-research.html

"Pennsylvania Assembly: Reply to the Governor, 11 November 1755," *Votes and*

Proceedings of the House of Representatives. 1755–1756 (1756). pp. 19–21.
https://founders.archives.gov/documents/Franklin/01-06-02-0107

Perry, Jane. "Breaches of Safety Regulations are Probable Cause of Recent
SARS Outbreak, WHO Says." *BMJ*. 328 (7450): 1222. May 25, 2004. https://
www.ncbi.nlm.nih.gov/pmc/articles/PMC416634/

Pham, Sherisse. "How Much Has the US Lost from China's IP Theft?" *CNN*.
March 23, 2018. https://money.cnn.com/2018/03/23/technology/china-us-
trump-tariffs-ip-theft/index.html

Potts, Jeremy and Alex, Dan. "USS Theodore Roosevelt (CVN-71)." *Military Factory*. April 13, 2020. https://www.militaryfactory.com/ships/detail.asp?
ship_id=USS-Theodore-Roosevelt-CVN71

Power Team, China. "How is China Feeding Its Population of 1.4 Billion?" *China
Power*. January 25, 2017. https://chinapower.csis.org/china-food-security/

Pradhan, Prashant, Pandey, Ashutosh Kumar, Mishra, Ashutosh Kumar, et al.
"Uncanny Similarity of Unique Inserts in the 2019-nCoV Spike Protein to
HIV-1 gp120 and Gag." *bioRxiv*. accessed on June 21, 2022. https://www.
biorxiv.org/content/10.1101/2020.01.30.927871v1

Pray, Leslie A. "Discovery of DNA Structure and Function: Watson and Crick."
Nature.2008.https://www.nature.com/scitable/topicpage/discovery-of-dna-
structure-and-function-watson-397/

"President Trump with Coronavirus Task Force Briefing." *C-SPAN*. March 9,
2020. 33:53. https://www.c-span.org/video/?470172-1/president-trump-
coronavirus-task-force-briefing

Press, Associated. "U.S. Moves to Drop Visa Fraud Charge Against a Chinese
Researcher." *NPR*. July 23, 2021. https://www.npr.org/2021/07/23/1019680651/
us-china-visa-fraud-case-charges

"Proclamation on Suspension of Entry as Immigrants and Nonimmigrants of
Persons Who Pose a Risk of Transmitting 2019 Novel Coronavirus." *The
White House*. January 31, 2020. https://trumpwhitehouse.archives.gov/pres-
idential-actions/proclamation-suspension-entry-immigrants-
nonimmigrants-persons-pose-risk-transmitting-2019-novel-coronavirus/

PTI. "Trump's Move to Ban Travel Between US and China Not a Great Moment:
House Speaker Nancy Pelosi." *The New Indian Express*. April 27, 2020.
https://www.newindianexpress.com/world/2020/apr/27/trumps-move-to-
ban-travel-between-us-and-china-not-a-great-moment-house-speaker-
nancy-pelosi-2135693.html

Qiu, Jane. "How China's 'Bat Woman' Hunted Down Viruses from SARS to the
New Coronavirus." *Scientific American*. June 1, 2020. https://www.scientific
american.com/article/how-chinas-bat-woman-hunted-down-viruses-
from-sars-to-the-new-coronavirus1/

Quinn, Adam. "The Art of Declining Politely: Obama's Prudent Presidency and
the Waning of American Power." *International Affairs*. Vol. 87. No. 4. July
2011. https://www.jstor.org/stable/20869760

Rado, Alicia Di. "Since Ancient Times, Biological Weapons Have Been Part of

Man's Arsenal." *USC News*. April 25, 2003. https://news.usc.edu/1872/Since-ancient-times-biological-weapons-have-been-part-of-man-s-arsenal/

Ratcliffe, John. "China is National Security Threat No. 1." *Wall Street Journal*. December 3, 2020. https://www.wsj.com/articles/china-is-national-security-threat-no-1-11607019599

Ratcliffe, John. "John Ratcliffe: China Olympics 2022 – COVID Cover Up By Country's Leaders Means They Should Forfeit Games." *Fox News*. August 2, 2021. https://www.foxnews.com/opinion/china-olympics-2022-covid-cover-up-games-john-ratcliffe

Re, Gregg. "Coronavirus Timeline Shows Politicians', Media's Changing Rhetoric on Risk of Pandemic." *Fox News*. April 20, 2020. https://www.foxnews.com/politics/from-new-york-to-canada-to-the-white-house-initial-coronavirus-responses-havent-aged-well

"Reconstruction of the 1918 Influenza Pandemic Virus." *CDC*. accessed on June 21, 2022. https://www.cdc.gov/flu/about/qa/1918flupandemic.htm

Regalado, Antonio. "China's CRISPR Twins Might Have Had Their Brains Inadvertently Enhanced." *Technology Review*. February 21, 2019. https://www.technologyreview.com/s/612997/the-crispr-twins-had-their-brains-altered/

Regalado, Antonio. "Pet Cloning is Bringing Human Cloning a Little Bit Closer." *Technology Review*. April 13, 2018. https://www.technologyreview.com/2018/04/13/143901/human-cloning-just-got-a-little-bit-closer-heres-why/

Reingold, Olivia. "Pompeo Insists COVID-19 Leaked from a Chinese Lab." *Politico*. June 13, 2021. https://www.politico.com/news/2021/06/13/pompeo-covid-chinese-lab-493986

"Remarks Before the Nazi War Criminals Interagency Working Group." *National Archives*. accessed on March 25, 2021. https://www.archives.gov/iwg/research-papers/weitzman-remarks-june-1999.html

"Research, Teaching, and Service for Health in China." *UNC Project-China*. accessed on May 4, 2020. https://www.med.unc.edu/medicine/infdis/china/

Rhew, David. "Successful COVID-19 Vaccine Delivery Requires Strong Tech Partnerships." *Official Microsoft Blog*. December 11, 2020. https://blogs.microsoft.com/blog/2020/12/11/successful-covid-19-vaccine-delivery-r requires-strong-tech-partnerships/

"Ribosome." *National Human Genome Research Institute*. June 16, 2022. https://www.genome.gov/genetics-glossary/Ribosome

Riotta, Chris. "From HIV to Covid-19: Fauci on His 'Complicated Relationship' with Activist Larry Kramer." *NBC*. October 2, 2020. https://www.nbcnews.com/feature/nbc-out/hiv-covid-19-dr-fauci-his-complicated-relationship-larry-kramer-n1241684

Roberts, Ken. "China More Dominant Than Ever in Covid-Related 'PPE' – And U.S. Flags." *Forbes*. September 19, 2020. https://www.forbes.com/sites/kenroberts/2020/09/19/china-more-dominant-than-ever-in-covid-related-ppe---and-us-flags/?sh=35444d6117f7

Rodrigues, Savio. "COVID-19 Secret is with Baric, Daszak, and Zhengli." *Sunday Guardian Live*, September 11, 2021. https://www.sundayguardianlive.com/news/covid-19-secret-baric-daszak-zhengli

Roelcke, Volker. "Nazi Medicine and Research on Human Beings." *The Lancet.* December 2004. https://www.thelancet.com/journals/lancet/article/PIIS0140-6736(04)17619-8/fulltext

Roffey, R., Tegnell, A., and Elgh, F. "Biological Warfare in a Historical Perspective." *Clinical Microbiology and Infection.* Volume 8, Issue 8. August 2002, pp. 450–54. https://www.sciencedirect.com/science/article/pii/S1198743 X14626343

Rogin, Josh. "State Department Cables Warned of Safety Issues In the Wuhan Lab Studying Bat Coronaviruses." *Washington Post.* April 14, 2020. https://www.washingtonpost.com/opinions/2020/04/14/state-department-cables-warned-s safety-issues-wuhan-lab-studying-bat-coronaviruses/?fbclid=IwAR3B7IYoB7d1ZQK28Y74IXTshFBPEwl5sZxDmwlMqSF4wWZ--TMZaIxtnqE

Rubik, Beverly and Brown, Robert R. "Evidence for a Connection Between Coronavirus Disease-19 and Exposure to Radiofrequency Radiation from Wireless Communications Including 5G." *Pub Med.* October 26, 2021. 7(5): 666–681. https://www.ncbi.nlm.nih.gov/pmc/articles/PMC8580522/

Rubin, Michael. "Why Tom Cotton and Others Are Right to Question Where Coronavirus Started." *The National Interest.* March 20, 2020. https://nationalinterest.org/blog/buzz/why-tom-cotton-and-others-are-right-question-where-coronavirus-started-135242

Sachs, Jeffrey D. and Harrison, Neil L. "Questions Surrounding the Origins of COVID-19 Remain Unanswered." *Boston Globe.* May 31, 2022. https://www.bostonglobe.com/2022/05/31/opinion/questions-surrounding-origins-covid-19-remain-unanswered/

"Samuel Slater: American Factory System." *PBS.* accessed on March 7, 2021. https://www.pbs.org/wgbh/theymadeamerica/whomade/slater_hi.html

Schmidt, Eric, Work, Bob, et al. "Final Report: National Security Commission on Artificial Intelligence." *National Security Commission on Artificial Intelligence.* February 2021. https://www.nscai.gov/wp-content/uploads/2021/03/Full-Report-Digital-1.pdf

"Scientists Are Working on HIV Vaccines Based on COVID Vaccine Tech." *Science Friday.* 17:08, April 1, 2022. https://www.sciencefriday.com/segments/fauci-hiv-vaccine-covid-mrna/

"Scientists, Politics, and Religion." *Pew Research Center.* July 9, 2009. https://www.pewresearch.org/politics/2009/07/09/section-4-scientists-politics-and-religion/

Scobey, Trevor, Yount, Boyd L., Sims, Amy C., and Baric, Ralph S. "Reverse Genetics with a Full-Length Infectious cDNA of the Middle East Respiratory Syndrome Coronavirus." *PNAS.* August 13, 2013. https://www.pnas.org/doi/10.1073/pnas.1311542110

BIBLIOGRAPHY

Seedhouse, Erik. "The Human Clone Market." *Beyond Human.* August 2, 2014, pp. 51–64. https://www.ncbi.nlm.nih.gov/pmc/articles/PMC7122979/

Sharma, Vaishali Basu. "Sars-cov2 is a Chimera with HIV Gene Manipulation." *PPF.* accessed on June 21, 2022. https://ppf.org.in/opinion/sars-cov2-is-a-chimera-with-hiv-gene-manipulation

Shepardson, David. "Chinese Genetics Company BGI Denies U.S. Human Rights Accusations." *Reuters.* July 21, 2020. https://www.reuters.com/article/us-usa-china-human-r rights/chinese-genetics-company-bgi-denies-u-s-human-rights-accusations-idUSKCN24N00A

Shibo, Zhang. *New Highland of War* (Beijing: National University Press, 2017).

Shoham, Dany. "About the Genomic Origin and Direct Source of the Pandemic Virus." *Journal of Defence Studies.* Vol. 16. No. 2. April–June 2022. pp. 79–92. https://www.idsa.in/system/files/jds/jds-16-2-2022-dany-shoham_compressed.pdf

Shoham, Dany. "China's Biological Warfare Programme: An Integrative Study with Special Reference to Biological Weapons Capabilities." *Journal of Defence Studies.* Vol. 9. No. 2. April 2015. https://idsa.in/jds/9_2_2015_Chinas BiologicalWarfareProgramme

Siddiki, Mahbube K. "China as the World Leader in Nanotechnology: Another Wakeup Call for the West." *Small Wars Journal.* March 12, 2022. https://smallwarsjournal.com/jrnl/art/china-world-leader-nanotechnology-another-wakeup-call-west

Sigalos, MacKenzie. "You Can't Sue Pfizer or Moderna If You Have Severe Covid Vaccine Side Effects. The Government Likely Won't Compensate You for Damages Either." *CNBC.* December 23, 2020. https://www.cnbc.com/2020/12/16/covid-vaccine-side-effects-compensation-lawsuit.html

Smith, Elliot. "China's Surging Food Prices Won't Weaken Its Hand in the Trade War, Economists Say." *CNBC.* August 9, 2019. https://www.cnbc.com/2019/08/09/chinas-surging-food-prices-wont-weaken-its-hand-in-the-trade-war.html

Spalding, Robert. *Stealth War: How China Took Over While America's Elite Slept* (New York: Penguin, 2019).

"Spotlight on Wuhan." *Nature.* May 13, 2015. https://www.ncbi.nlm.nih.gov/pmc/articles/PMC7095288/

Statement by President Joe Biden on the Investigation into the Origins of COVID-19." *The White House.* May 26, 2021. https://www.whitehouse.gov/briefing-room/statements-releases/2021/05/26/statement-by-president-joe-biden-on-the-investigation-into-the-origins-of-covid-19/

Subbaraman, Nidhi. "Scientists' Fears of Racial Bias Surge Amid US Crackdown on China Ties." *Nature.* October 29, 2021. https://www.nature.com/articles/d41586-021-02976-8

Taylor, Adam, Rauhala, Emily, and Sorensen, Martin Selsoe. "In New Documentary, WHO Scientist Says Chinese Officials Pressured Investigation to

Drop Lab-Leak Hypothesis." *Washington Post*. August 12, 2021. https://www.washingtonpost.com/world/2021/08/12/who-origins-embarek/

Tenchov, Rumina. "Understanding the Nanotechnology in COVID-19 Vaccines." *CAS*. February 18, 2021. https://www.cas.org/resource/blog/understanding-nanotechnology-covid-19-vaccines

Thacker, Paul D. "COVID-19: Lancet Investigation Into Origin of Pandemic Shuts Down Over Bias Risk." *BMJ*. 2021. 375:n2414. https://www.bmj.com/content/375/bmj.n2414

"The China Threat: Chinese Talent Plans Encourage Trade Secret Theft, Economic Espionage." *Federal Bureau of Investigation*. accessed on June 29, 2022. https://www.fbi.gov/investigate/counterintelligence/the-china-threat/chinese-talent-plans

"The Chinese Consulate in Houston May Have Provided Financial and Logistical Support to Protest Groups," *The Liberty Web*, 8 September 2020. https://eng.the-liberty.com/2020/8025/

"'The Code Breaker': Jennifer Doudna and How CRISPR May Revolutionize Mankind." *CBS*. March 7, 2021. https://www.cbsnews.com/news/crispr-jennifer-doudna-walter-isaacson-the-code-breaker/

"The Future Trend of the World's New Military Revolution." *Reference News*. August 24, 2017. https://web.archive.org/web/20190823210313/http://www.xinhuanet.com/politics/2017-08/24/c_129687890.htm

"The Human Genome Project." *National Human Genome Research Institute*. accessed on June 15, 2022. https://www.genome.gov/human-genome-project

"The Institute of Human Virology: 2019 Annual Report." *The University of Maryland School of Medicine*. p. 14. accessed on June 26, 2022. https://www.ihv.org/media/SOM/Microsites/IHV/documents/Annual-Report/2019-IHV-Annual-Report-final-v7-low-res.pdf

"The Newest Members of Beijing's Police Force are 6 Cloned K9s." *CBS News*. November 21, 2019. https://www.cbsnews.com/news/cloned-police-dogs-china-beijing-security-forces/

"The Rosenberg Trial." *Atomic Heritage Foundation*. April 25, 2018. https://www.atomicheritage.org/history/rosenberg-trial

"The Spy Next Door?" *Ash Clinical News*. June 2019. https://ashpublications.org/ashclinicalnews/news/4599/The-Spy-Next-Door

"Tianxia (All Under Heaven)." *Chinese Thought*. accessed on July 1, 2022. https://www.chinesethought.cn/EN/shuyu_show.aspx?shuyu_id=2161

Times, Hindustan. "U.S. Convicts Harvard Prof for Lying About China Ties. Crackdown to Curb Beijing Influence in U.S.?" *YouTube*. December 22, 2021. https://www.youtube.com/watch?v=Igo8_wOrm6c

Today, NTD News. "China Pursues 'Brain Control' Weaponry." *NTD*. January 11, 2022. https://www.ntd.com/china-pursues-brain-control-weaponry_726017.html

Towey, Robert. "Fauci Says Rand Paul is 'Egregiously Incorrect' About Gain of

BIBLIOGRAPHY

Function Research in Senate Showdown." *CNBC*, November 4, 2021. https://www.cnbc.com/2021/11/04/fauci-says-rand-paul-egregiously-incorrect-about-gain-of-function-research.html

"Transcript: Matt Pottinger on 'Face the Nation,' February 21, 2021." *CBS News*. February 21, 2021. https://www.cbsnews.com/news/transcript-matt-pottinger-on-face-the-nation-february-21-2021/

Truex, Rory. "What the Fear of China is Doing to American Science." *The Atlantic*. February 16, 2021. https://www.theatlantic.com/ideas/archive/2021/02/fears-about-china-are-disrupting-american-science/618031/

"Trump Accuses China of 'Raping' US with Unfair Trade Policy." *BBC*. May 2, 2016. https://www.bbc.com/news/election-us-2016-36185012

"Trump-Taiwan Call Breaks US Policy Stance." *BBC*. December 3, 2016. https://www.bbc.com/news/world-us-canada-38191711

"UNC Baric 12 30 21." *U.S. Right to Know*. accessed on July 6, 2022. https://usrtk.org/wp-content/uploads/2021/12/UNC_Baric_12.30.21.pdf

"Understanding mRNA COVID Vaccines." *Centers for Disease Control*. January 4, 2022. https://www.cdc.gov/coronavirus/2019-ncov/vaccines/different-vaccines/mrna.html

"University Researcher Sentenced to Prison for Lying on Grant Applications to Develop Scientific Expertise for China." *Department of Justice*. May 14, 2021. https://www.justice.gov/opa/pr/university-researcher-sentenced-prison-lying-grant-applications-develop-scientific-expertise

"U of G Chemists Find Microwaves May Help Treat COVID-19." *University of Guelph News*. February 28, 2022. https://news.uoguelph.ca/2022/02/u-of-g-chemists-find-microwaves-may-help-treat-covid-19/

"U.S.S. Theodore Roosevelt." *U.S. Navy*. September 19, 2001. https://www.wsfa.com/story/477688/uss-theodore-roosevelt/

Viglione, Giuliana. "Scientists Strongly Back Joe Biden for US President in Nature Poll." *Nature*. October 23, 2020. https://www.nature.com/articles/d41586-020-02963-5

Vogel, Emily S., Anderson, Monica, Porteus, Margaret, et al. "Americans and 'Cancel Culture': Where Some See Calls for Accountability, Others See Censorship, Punishment." *Pew Research Center*. May 19, 2021. https://www.pewresearch.org/internet/2021/05/19/americans-and-cancel-culture-where-some-see-calls-for-accountability-others-see-censorship-punishment/

Walker, James. "Mike Pompeo Claims Intel Officials 'Didn't Want to Talk' About Wuhan Lab Leak Theory." *Newsweek*. June 4, 2021. https://www.newsweek.com/mike-pompeo-intel-officials-wuhan-lab-leak-cover-1597494

Walsh, Bryan. "The World Is Not Ready for the Next Pandemic." *Time*. May 4, 2017. https://time.com/magazine/us/4766607/may-15th-2017-vol-189-no-18-u-s/

Wang, Dan. "China Hawks Don't Understand How Science Advances," *The Atlantic*. December 12, 2021. https://www.theatlantic.com/ideas/

archive/2021/12/china-initiative-intellectual-property-theft/
621058/

Washington, Associated Press in. "US Government Hack Stole Fingerprints of
5.6 Million Federal Employees." *The Guardian*. September 23, 2015. https://
www.theguardian.com/technology/2015/sep/23/us-government-hack-
stole-fingerprints

Washington, Reuters in. "China the 'Greatest Threat to Democracy and Freedom,' U.S. Spy Chief Warns." *The Guardian*. December 3, 2020. https://www.
theguardian.com/us-news/2020/dec/03/china-beijing-america-
democracy-freedom

Weatherby, Todd and Jonakin, Kelli. "Executive Conversations: Accelerating
COVID-19 Vaccine Development with Marcello Damiani, Chief Digital and
Operational Excellence Officer at Moderna." *AWS*. March 1, 2021. https://
aws.amazon.com/blogs/industries/executive-conversations-accelerating-
covid-19-vaccine-development-with-marcello-damiani-chief-digital-and-
operational-excellence-officer-at-moderna/

Web Desk, Fox40. "A Look Back: Stuck on Cruise Ships, Local Couples Relied on
Wavering Hope as COVID-19 Became a Shocking Reality." *Fox 40*. March 15,
2021. https://fox40.com/news/local-news/a-look-back-stuck-on-cruise-
ships-local-couples-relied-on-wavering-hope-as-covid-19-became-a-
shocking-reality/

Weber, Noah. "In Taiwan, Is It 'COVID-19' or 'Wuhan Pneumonia'?" *Sup China*.
April 6, 2020. https://supchina.com/2020/04/06/in-taiwan-is-it-covid-19-
or-wuhan-pneumonia/

Wehner, Mike. "Scientists Create World's First Pig-Monkey Hybrid in China."
New York Post. December 9, 2019. https://nypost.com/2019/12/09/
scientists-create-worlds-first-pig-monkey-hybrid-in-china/

Weichert, Brandon J. "China is Now Under 'War-Time Controls'." *The Weichert
Report*. February 17, 2020. https://theweichertreport.wordpress.com/2020/
02/17/china-is-now-under-war-time-controls/

Weichert, Brandon J. "China Might Try to Take Taiwan." *American Greatness*.
April 18, 2020. https://amgreatness.com/2020/04/18/china-might-try-
to-take-taiwan/

Weichert, Brandon J. "China's Growing Biotech Threat." *New English Review*.
February 2020. https://www.newenglishreview.org/articles/chinas-
growing-biotech-threat/

Weichert, Brandon J. "How Covid Made the Present World." *Asia Times*. June 23,
2021. https://asiatimes.com/2021/06/how-covid-made-the-present-world/

Weichert, Brandon J. "Much More Than a Trade War with China." *New English
Review*. June 2019. https://www.newenglishreview.org/articles/much-more-
than-a-trade-war-with-china/

Weichert, Brandon J. "China's Quest for Exotic Tech: Brain Control Interface."
The Weichert Report. February 7, 2021. https://theweichertreport.wordpress.
com/2021/02/07/chinas-quest-for-exotic-tech-brain-control-interface/

Weichert, Brandon J. "Understanding Digital Security." *Real Clear Public Affairs.* August 19, 2019. https://www.realclearpublicaffairs.com/articles/2019/08/19/understanding_digital_securit y_18786.html

Weichert, Brandon J. *Winning Space: How America Remains a Superpower* (Alexandria: Republic Book Publishers, 2020), p. 59.

Weichert, Brandon J. "Xi Jinping's Power Struggle Shakes the World." *Washington Times.* May 25, 2022. https://www.washingtontimes.com/news/2022/may/25/xi-jinpings-power-struggle-shakes-the-world/

Weixel, Nathaniel. "Federal Stockpile of Emergency Medical Equipment Depleted, House Panel Says." *The Hill.* April 8, 2020. https://thehill.com/policy/healthcare/491871-federal-stockpile-of-emergency-medical-equipment-depleted-house-panel-says/

"What is Nanotechnology?" *National Nanotechnology Institute.* accessed on June 30, 2022. https://www.nano.gov/nanotech-101/what/definition

Wong, Edward. "A Chinese Empire Reborn." *New York Times.* January 5, 2018. https://www.nytimes.com/2018/01/05/sunday-review/china-military-economic-power.html

Wu, Canrong, Zheng, Mengzhu, Yang, Yueying, et al. "Furin: A Potential Therapeutic Target for COVID-19," *iScience.* Volume 23. Issue 10. October 2020. https://www.sciencedirect.com/science/article/pii/S2589004220308348#!

Xiao, Botao and Xiao, Lei. "The Possible Origins of 2019-nCoV Coronavirus." *South China University of Technology.* February 6, 2020. https://s.rfi.fr/media/display/2f7b52c0-87b0-11ea-b8a0-005056bff430/Xiao%20Botao%20-%20The%20possible%20origins%20of%20the%202019-nCoV%20virus.pdf

Xie, John. "Chinese Lab with Checkered Safety Record Draws Scrutiny Over COVID-19." *Voice of America.* April 21, 2020. https://www.voanews.com/a/covid-19-pandemic_chinese-lab-checkered-safety-record-draws-scrutiny-over-covid-19/6187947.html

"Xi Calls for Developing China into World Science and Technology Leader." *en.people.cn.* May 29, 2018. http://en.people.cn/n3/2018/0529/c90000-9464968.html?fbclid=IwAR2HDaRp9GXqob8nwgUJgaGIxFXD8-23nEtV1asqYfm-FkTONkJ__tz9MIk

Yang, Jianli and Rhodes, Aaron. "How China Has Crushed Hong Kong's Democracy." *National Review.* March 18, 2021. https://www.nationalreview.com/2021/03/how-china-has-crushed-hong-kongs-democracy/

Yang, Lin. "Pandemic Exposes Perils of Global Reliance on China for Drug Supplies." *VOA.* May 19, 2020. https://www.voanews.com/a/science-health_pandemic-exposes-perils-global-reliance-china-drug-supplies/6189571.html

Yegorov, Oleg. "Why Did the Rosenbergs Spy and Die for Communism?" *Russia Beyond.* March 16, 2018. https://www.rbth.com/history/327844-why-did-the-rosenbergs-spy-for-the-ussr

Young, Alison. "Deleted COVID-19 Genetic Fingerprints Show It's Still Possible to Dig for Lab Leak Evidence." *USA Today.* June 24, 2021. https://www.usa today.com/story/opinion/2021/06/24/covid-19-lab-leak-investigation-deleted-genetic-fingerprints-show-its-still-possible-dig-lab-leak-ev/7778194002/

Young, Alison and Blake, Jessica. "Near Misses at UNC Chapel Hill's High Security Lab Illustrate Risk of Accidents with Coronaviruses." *ProPublica.* August 17, 2020. https://www.propublica.org/article/near-misses-at-unc-chapel-hills-high-security-lab-illustrate-risk-of-accidents-with-coronaviruses

Zha, Qiang. "Can China Reverse the Brain Drain?" *University World News.* March 28, 2014. https://www.universityworldnews.com/post.php?story=20140326132305490

Zhang, Wendong, Rodriguez, Lulu, Qu, Shuyang. "3 Reasons Midwest Farmers Hurt by the U.S.-China Trade War Still Support Trump." *The Conversation.* November 4, 2019. https://theconversation.com/3-reasons-midwest-farmers-hurt-by-the-u-s-china-trade-war-still-support-trump-126303

Zompa, Tenzin. "Chinese Military's Epidemiologist Worked with Fired Scientist at Canada's Top Disease Lab: Report." *The Print.* September 17, 2021. https://theprint.in/world/chinese-militarys-epidemiologist-worked-with-fired-scientist-at-canadas-top-disease-lab-report/734940/

Zong, Zhi, Wei,Yujun, Ren, Jiang, et al. "The Intersection of COVID-19 and Cancer: Signaling Pathways and Treatment Implications." *Molecular Cancer.* 20. 76. 2021. https://molecular-cancer.biomedcentral.com/articles/10.1186/s12943-021-01363-1

FOREWORD

1 Gordon G. Chang is the author of *The Coming Collapse of China* (Random House). Follow him on Twitter @GordonGChang.

2 "US Intel Community Remains 'Divided' on COVID-19 Origins," *Al Jazeera*, 27 August 2021. https://www.aljazeera.com/news/2021/8/27/us-intel-community-remains-divided-on-covid-19-origin

INTRODUCTION

1 "China: New Hong Kong Law a Roadmap for Repression," *Human Rights Watch*, 29 July 2020, https://www.hrw.org/news/2020/07/29/china-new-hong-kong-law-roadmap-repression

2 "Laogai Research Foundation," accessed on 24 September 2022. https://laogairesearch.org/laogai-system/

3 Sonia Elks, "China is Harvesting Organs from Falun Gong Members, Finds Expert Panel," *Reuters*, 17 June 2019. https://www.reuters.com/article/us-britain-china-rights/china-is-harvesting-organs-from-falun-gong-members-finds-expert-panel-idUSKCN1TI236

4 Paul Mazor and Aaron Krolik, "A Surveillance Net Blankets China's Cities, Giving Police Vast Powers," *New York Times*, 17 December 2019. https://www.nytimes.com/2019/12/17/technology/china-surveillance.html

5 Beina Xu and Eleanor Albert, "Media Censorship in China," *Council on Foreign Relations*, 17 February 2017. https://www.cfr.org/backgrounder/media-censorship-china

6 Sherisse Pham, "How Much Has the US Lost from China's IP Theft?" *CNN*, 23 March 2018. https://money.cnn.com/2018/03/23/technology/china-us-trump-tariffs-ip-theft/index.html

7 Mara Hvistendahl, "Exclusive: Major U.S. Cancer Center Ousts 'Asian' Researchers After NIH Flags Their Foreign Ties," *Science Magazine*, 19 April

2019.https://www.sciencemag.org/news/2019/04/exclusive-major-us-cancer-center-ousts-asian-researchers-after-nih-flags-their-foreign

8 "Harvard University Professor and Two Chinese Nationals Charged in Three Separate China Related Cases," *Department of Justice*, 28 January 2020. https://www.justice.gov/opa/pr/harvard-university-professor-and-two-chinese-nationals-charged-three-separate-china-related

9 Zhang Shibo, *New Highland of War* (Beijing: National University Press, 2017).

10 "Xi Calls for Developing China into World Science and Technology Leader," *en.people.cn*, 29 May 2018. http://en.people.cn/n3/2018/0529/c90000-9464968.html?fbclid=IwAR2HDaRp9GXqob8nwgUJgaGIxFXD8-23nEtV1asqYfm-FkTONkJ__tz9MIk

11 Jennifer Manongdo, "China Could Harvest US Athletes' DNA at Winter Olympics, Lawmaker Warns," *International Business Times*, 16 June 2021. https://www.ibtimes.com/china-could-harvest-us-athletes-dna-beijing-winter-olympics-lawmaker-warns-3227033

12 "Biotech: What's Driving China's Biotechnology Revolution?" *The Motley Fool*, 26 December 2018. https://www.youtube.com/watch?v=EemvGxof4Ro&feature=emb_logo

13 Hadi Khatib, "Cloning Pets and Animals Now Common Practice. Is Cloning Humans Next?" *AME Info*, 27 August 2019. https://www.ameinfo.com/indutry/healthcare/cloning-pets-and-animals-now-common-practice-is-cloning-humans-next

14 Mike Wehner, "Scientists Create World's First Pig-Monkey Hybrid in China," *New York Post*, 9 December 2019. https://nypost.com/2019/12/09/scientists-create-worlds-first-pig-monkey-hybrid-in-china/

15 Brandon J. Weichert, *Winning Space: How America Remains a Superpower* (Alexandria: Republic Book Publishers, 2020), p. 59.

16 Robert Spalding, *Stealth War: How China Took Over While America's Elite Slept* (New York: Penguin, 2019), pp. 204–06.

17 "Harvard University Professor and Two Chinese Nationals Charged in Three Separate China Related Cases," *United States Department of Justice*, 28 January 2020. https://www.justice.gov/opa/pr/harvard-university-professor-and-two-chinese-nationals-charged-three-separate-china-related

18 Kai-Fu Lee, *AI Superpowers: China, Silicon Valley, and the New World Order* (New York: Houghton Mifflin Harcourt, 2018), pp. 15–16.

19 Dennis Normile, "CRISPR Bombshell: Chinese Researcher Claims to Have Gene-Edited Twins," *Science Magazine*, 26 November 2018. https://www.sciencemag.org/news/2018/11/crispr-bombshell-chinese-researcher-claims-have-created-gene-edited-twins

20 Antonio Regalado, "China's CRISPR Twins Might Have Had Their Brains Inadvertently Enhanced," *Technology Review*, 21 February 2019. https://www.technologyreview.com/s/612997/the-crispr-twins-had-their-brains-altered/

21 Dennis Normile, "Chinese Scientists Who Produced Genetically Altered Babies Sentenced to 3 Years in Jail," *Science Magazine*, 30 December 2019. https://www.sciencemag.org/news/2019/12/chinese-scientist-who-produced-genetically-altered-babies-sentenced-3-years-jail

22 Steven W. Mosher, "Don't Buy China's Story: The Coronavirus May Have Leaked from a Lab," *New York Post*, 22 February 2020. https://nypost.com/2020/02/22/dont-buy-chinas-story-the-coronavirus-may-have-leaked-from-a-lab/

23 Michael Rubin, "Why Tom Cotton and Others Are Right to Question Where Coronavirus Started," *The National Interest*, 20 March 2020. https://nationalinterest.org/blog/buzz/why-tom-cotton-and-others-are-right-question-where-coronavirus-started-135242

24 Steven W. Mosher, "China Must Release the Secret Records of the Wuhan Biolabs," *American Greatness*, 27 March 2020. https://amgreatness.com/2020/03/27/china-must-release-the-secret-records-of-the-wuhan-biolabs/

25 Morgan Chalfant, "Trump's National Security Adviser Says China 'Covered Up,' Coronavirus," *The Hill*, 11 March 2020. https://thehill.com/homenews/administration/487011-trumps-national-security-adviser-says-china-covered-up-coronavirus

26 "Research, Teaching, and Service for Health in China," *UNC Project-China*, accessed on 4 May 2020. https://www.med.unc.edu/medicine/infdis/china/

27 Todd Ackerman, "MD Anderson Ousts 3 Scientists Over Concerns About Chinese Conflicts of Interest," *The Houston Chronicle*, 20 April 2020. https://www.houstonchronicle.com/news/houston-texas/houston/article/MD-Anderson-fires-3-scientists-over-concerns-13780570.php

28 Nuala Moran, "Researchers Trace COVID-19's Family Tree to Battle Outbreak and 'Infodemic,'" *Bio World*, 14 February 2020. https://www.bioworld.com/articles/433087-article-headline

29 "Samuel Slater: American Factory System," *PBS*, accessed on 7 March 2021. https://www.pbs.org/wgbh/theymadeamerica/whomade/slater_hi.html

30 Ibid.

31 Gordon G. Chang, *The Great U.S.-China Tech War* (New York: Encounter Books, 2020), pp. 1–14.

32 Celia Chen, "US-China Tech War: Beijing's Main Policy Lender Pledges US $62 Billion to Fund Tech Innovation," *South China Morning Post*, 4 March 2021. https://www.scmp.com/tech/policy/article/3124094/us-china-tech-war-beijings-main-policy-lender-pledges-us62-billion-fund

33 Elsa B. Kania and Wilson Vorndick, "Weaponizing Biotech: How China's Military is Preparing for a 'New Domain of Warfare,'" *Defense One*, 14 August 2019. https://www.defenseone.com/ideas/2019/08/chinas-military-pursuing-biotech/159167/

34 Ken Dilanian, "China Has Done Human Testing to Create Biologically Enhanced Super Soldiers, Says Top U.S. Official," *NBC News*, 3 December 2020. https://www.nbcnews.com/politics/national-security/china-has-

done-human-testing-create-biologically-enhanced-super-soldiers-n1249914

35 Arjun Kharpal, "China's Baidu is in Talks to Raise Up to $2 Billion to Launch a Stand-Alone Biotech Company," *CNBC*, 10 September 2020. https://www.cnbc.com/2020/09/10/baidu-raising-money-for-biotech-firm-that-uses-artificial-intelligence-.html

36 Kirsty Needham, "Special Report: COVID Opens New Doors for China's Gene Giant," *Reuters*, 5 November 2020. https://www.reuters.com/article/us-health-coronavirus-bgi-specialreport/special-report-covid-opens-new-doors-for-chinas-gene-giant-idUSKCN2511CE

37 Elsa B. Kania and Wilson Vorndick, "China's Military Biotech Frontier: CRISPR, Military-Civil Fusion, and the New Revolution in Military Affairs," *Jamestown Foundation*, Vol. 19, Issue 18, 8 October 2019. https://jamestown.org/program/chinas-military-biotech-frontier-crispr-military-civil-fusion-and-the-new-revolution-in-military-affairs/

38 "China: Minority Region Collects DNA from Millions," *Human Rights Watch*, 13 December 2017. https://www.hrw.org/news/2017/12/13/china-minority-region-collects-dna-millions#

39 David Shepardson, "Chinese Genetics Company BGI Denies U.S. Human Rights Accusations," *Reuters*, 21 July 2020. https://www.reuters.com/article/us-usa-china-human-rights/chinese-genetics-company-bgi-denies-u-s-human-rights-accusations-idUSKCN24N00A

40 "Beijing Genomics Institute (BGI) Forms Children's Disease Partnership with Children's Hospital of Philadelphia," *BioSpace*, 18 November 2011. https://www.biospace.com/article/releases/beijing-genomics-institute-bgi-forms-children-s-disease-partnership-with-children-s-hospital-of-philadelphia-/

41 Kirsty Needham, "Exclusive: China Gene Firm Providing Worldwide COVID Tests Worked with Chinese Military," *Reuters*, 30 January 2021. https://www.reuters.com/article/us-china-genomics-military-exclusive/exclusive-china-gene-firm-providing-worldwide-covid-tests-worked-with-chinese-military-idUSKBN29Z0HA

42 Douglas McGray, "Biotech's Black Market," *Mother Jones*, September/October 2002. https://www.motherjones.com/politics/2002/09/biotechs-black-market/

43 Jonathan Cheng, "China is the Only Major Economy to Report Economic Growth for 2020," *Wall Street Journal*, 18 January 2021. https://www.wsj.com/articles/china-is-the-only-major-economy-to-report-economic-growth-for-2020-11610936187

44 Ashley Parker, Josh Dawsey, Matt Viser, and Michael Scherer, "How Trump's Erratic Behavior and Failure on Coronavirus Doomed His Reelection," *Washington Post*, 7 November 2020. https://www.washingtonpost.com/elections/interactive/2020/trump-pandemic-coronavirus-election/

NOTES

CHAPTER 1

1   Remco Zwetsloot, Jack Corrigan, Emily S. Weinstein, Dahlia Petersen, Diana Gehlhaus, and Ryan Feusiak, "China is Fast Outpacing U.S. STEM PhD Growth," *Center for Security and Emerging Technology*, August 2021. https://cset.georgetown.edu/publication/china-is-fast-outpacing-u-s-stem-phd-growth/

CHAPTER 2

1   Kyle Mizokami, "Here's Every Aircraft Carrier in the World," *Popular Mechanics*, 16 December 2020. https://www.popularmechanics.com/military/navy-ships/g2412/a-global-roundup-of-aircraft-carriers/

2   "U.S.S. Theodore Roosevelt," *U.S. Navy*, 19 September 2001. https://www.wsfa.com/story/477688/uss-theodore-roosevelt/

3   Jeremy Potts and Dan Alex, "USS Theodore Roosevelt (CVN-71)," *Military Factory*, 13 April 2020. https://www.militaryfactory.com/ships/detail.asp?ship_id=USS-Theodore-Roosevelt-CVN71

4   Huntington Ingalls Industries. *Twitter.* 27 February 2015, 3:13 pm. https://twitter.com/HIIndustries/status/571402713201836032

5   "Timeline: Theodore Roosevelt COVID-19 Outbreak Investigation," *USNI News*, 23 June 2020. https://news.usni.org/2020/06/23/timeline-theodore-roosevelt-covid-19-outbreak-investigation

6   "Remarks Before the Nazi War Criminals Interagency Working Group," *National Archives*, accessed on 25 March 2021. https://www.archives.gov/iwg/research-papers/weitzman-remarks-june-1999.html

7   "Deng Xiaoping's '24-Character Strategy,'" *Global Security*, accessed on 25 March 2021. https://www.globalsecurity.org/military/world/china/24-character.htm

8   Sean McFate, "The New Rules of War: Victory in the Age of Durable Disorder," *The Westminster Institute*, 30 January 2019. https://westminster-institute.org/events/the-new-rules-of-war-victory-in-the-age-of-durable-disorder/

9   Andrew Court, "Both US Aircraft Carriers in the Pacific are Taken Out of Action for Up to a MONTH After Sailors Get Infected with Coronavirus – Giving China a Free Hand in the Region as the Pentagon Raises Threat Level to Second Highest Setting," *Daily Mail*, 27 March 2020. https://www.dailymail.co.uk/news/article-8161181/Sailors-aircraft-carriers-Pacific-coronavius.html

10 Brandon J. Weichert, "China Might Try to Take Taiwan," *American Greatness*, 18 April 2020. https://amgreatness.com/2020/04/18/china-might-try-to-take-taiwan/

NOTES

CHAPTER 3

1　"A Brief History of CRISPR-CAS-9 Genome-Editing Tools," *BiteSize Bio*, 30 June 2020. https://bitesizebio.com/47927/history-crispr/

2　Leslie A. Pray, "Discovery of DNA Structure and Function: Watson and Crick," *Nature*, 2008. https://www.nature.com/scitable/topicpage/discovery-of-dna-structure-and-function-watson-397/

3　Matthew Cobb, "Sexism in Science: Did Watson and Crick Really Steal Rosalind Franklin's Data?" *The Guardian*, 23 June 2015. https://www.theguardian.com/science/2015/jun/23/sexism-in-science-did-watson-and-crick-really-steal-rosalind-franklins-data

4　Alicia Di Rado, "Since Ancient Times, Biological Weapons Have Been Part of Man's Arsenal," *USC News*, 25 April 2003. https://news.usc.edu/1872/Since-ancient-times-biological-weapons-have-been-part-of-man-s-arsenal/

5　R. Roffey, A. Tegnell, and F. Elgh, "Biological Warfare in a Historical Perspective," *Clinical Microbiology and Infection*, Volume 8, Issue 8, August 2002, 450–54. https://www.sciencedirect.com/science/article/pii/S1198743X14626343

6　"1987: First Human Genetic Map," *National Human Genome Research Institute*, accessed on 15 June 2022. https://www.genome.gov/25520325/online-education-kit-1987-first-human-genetic-map

7　"The Human Genome Project," *National Human Genome Research Institute*, accessed on 15 June 2022. https://www.genome.gov/human-genome-project

8　Ibid.

9　"Ribosome," *National Human Genome Research Institute*, 16 June 2022. https://www.genome.gov/genetics-glossary/Ribosome

10　Anne Marie Helmenstine, "The Differences Between DNA and RNA," *ThoughtCo.*, 2 February 2020. https://www.thoughtco.com/dna-versus-rna-608191

11　Terry Gross, "CRISPR Scientist's Biography Explores Ethics of Rewriting the Code of Life," *NPR*, 8 March 2021. https://www.npr.org/transcripts/974751834

12　Reuters Fact Check, "Fact Check-mRNA Vaccines are Distinct from Gene Therapy, Which Alters Recipient's Genes," *Reuters*, 10 August 2021. https://www.reuters.com/article/factcheck-covid-mrna-gene/fact-check-mrna-vaccines-are-distinct-from-gene-therapy-which-alters-recipients-genes-idUSL1N2PH16N

13　"Big Pharma Exec: COVID Shots are 'Gene Therapy'" *Cause Action*, 15 November 2021. https://cqrcengage.com/causeaction/app/document/36532017

14　Walter Isaacson, "mRNA Technology Gave Us the First COVID-19 Vaccine.

It Could Also Upend the Drug Industry," *Time*, 11 January 2021. https://time.com/5927342/mrna-covid-vaccine/

15 "Understanding mRNA COVID Vaccines," *Centers for Disease Control*, 4 January 2022. https://www.cdc.gov/coronavirus/2019-ncov/vaccines/different-vaccines/mrna.html

16 Annalee Armstrong, "Big Pharma Partnerships, Record $22.7B Investment Raise Profile of Regenerative Medicine in 2021," *Fierce Biotech*, 5 April 2022. https://www.fiercebiotech.com/biotech/big-pharma-partnerships-record-investment-raise-profile-regenerative-medicine-2021

17 Brian Gormley, "Venture Investment in Crispr Gene Editing Spurs Innovation, Hunt for IP," *Wall Street Journal*, 10 March 2022. https://www.wsj.com/articles/venture-investment-in-crispr-gene-editing-spurs-innovation-hunt-for-ip-11646910000

18 Mathias Evers and Michael Chui, "The Promise and Peril of the Bio Revolution," *Project Syndicate*, 26 June 2021. https://www.project-syndicate.org/commentary/biological-innovation-promise-and-perils-by-matthias-evers-and-michael-chui-2021-01?h=bi2qIbNWmgTmtpEm2gn4nKlt51RPReGGwz9uiEwPpkY%3d&

19 Eva Frederick, "New CRISPR-Based Map Ties Every Human Gene to Its Function," *MIT News*, 9 June 2022. https://news.mit.edu/2022/crispr-based-map-ties-every-human-gene-to-its-function-0609

20 "'The Code Breaker': Jennifer Doudna and How CRISPR May Revolutionize Mankind," *CBS*, 7 March 2021. https://www.cbsnews.com/news/crispr-jennifer-doudna-walter-isaacson-the-code-breaker/

21 Ibid.

22 Hannah Kuchler, "Jennifer Doudna, Crispr Scientist, On the Ethics of Editing Humans," *Financial Times*, 31 January 2020. https://www.ft.com/content/6d063e48-4359-11ea-abea-0c7a29cd66fe

CHAPTER 4

1 Jane Mayer, "How Russia Helped Swing the Election for Trump," *New Yorker*, 24 September 2018. https://www.newyorker.com/magazine/2018/10/01/how-russia-helped-to-swing-the-election-for-trump

2 Richard Hanania, "It Isn't Your Imagination: Twitter Treats Conservatives More Harshly Than Liberals," *Quillette*, 12 February 2019. https://quillette.com/2019/02/12/it-isnt-your-imagination-twitter-treats-conservatives-more-harshly-than-liberals/

3 Mike Davis, "Big Tech Censorship of COVID Information Leads to Vaccine Hesitancy,"*Newsweek*,2November2021.https://www.newsweek.com/big-tech-censorship-covid-information-leads-vaccine-hesitancy-opinion-1644051

4 Ahmed Sule, "The 'COVID-Industrial Complex' – a Web of Big Pharma, Big Tech, and Politicians – Are Profiting Off the Pandemic at the Expense of the Public," *Business Insider*, 28 March 2021. https://www.businessinsider.

com/case-against-covid-industrial-complex-pandemic-pharma-politicians-tech-profiting-2021-3

5 Brandon J. Weichert, "China is Now Under 'War-Time Controls,'" *The Weichert Report*, 17 February 2020. https://theweichertreport.wordpress.com/2020/02/17/china-is-now-under-war-time-controls/

6 "Coronavirus: Welding Doors Shut," *CBC*, 27 February 2020. https://www.cbc.ca/player/play/1703503427818

7 "Coronavirus: Residents 'Welded' Inside Their Own Homes in China," *LBC*, 2 February 2020. https://www.lbc.co.uk/news/coronavirus-residents-welded-inside-their-own-home/

8 Katharina Hauck and Roshni Mehta, "The Economic Cost of China's Lasting Zero-COVID Strategy," *Think Global Health*, 28 April 2022. https://www.thinkglobalhealth.org/article/economic-cost-chinas-lasting-zero-covid-strategy

9 Christopher R. Berry, Anthony Fowler, Tamara Glazer, and Alec MacMillen, "Evaluating the Effects of Shelter-in-Place Policies During the COVID-19 Pandemic," *PNAS*, 25 March 2021. https://www.pnas.org/doi/10.1073/pnas.2019706118#T1

10 Brandon J. Weichert, "Xi Jinping's Power Struggle Shakes the World," *Washington Times*, 25 May 2022. https://www.washingtontimes.com/news/2022/may/25/xi-jinpings-power-struggle-shakes-the-world/

11 Laura Italiano, "Coronavirus 'Whistleblower' Nurse Says China Has 90,000 Sick," *New York Post*, 26 January 2020. https://nypost.com/2020/01/26/coronavirus-whistleblower-nurse-says-china-has-90000-sick/

12 "Li Wenliang: 'Wuhan Whistleblower' Remembered One Year On," *BBC*, 6 February 2021. https://www.bbc.com/news/world-asia-55963896

13 Fox40 Web Desk, "A Look Back: Stuck On Cruise Ships, Local Couples Relied On Wavering Hope as COVID-19 Became a Shocking Reality," *Fox 40*, 15 March 2021. https://fox40.com/news/local-news/a-look-back-stuck-on-cruise-ships-local-couples-relied-on-wavering-hope-as-covid-19-became-a-shocking-reality/

14 Alexandra Hutzler, "New Report Finds Cuomo's Controversial Nursing Home Guidance 'May Have Made a Bad Situation Worse,'" *Newsweek*, 19 February 2021. https://www.newsweek.com/new-report-finds-cuomos-controversial-nursing-home-guidance-may-have-made-bad-situation-worse-1570630

15 Post Editorial Board, "Federal Government Using Social-Media Giants to Censor Americans," *New York Post*, 6 September 2021. https://nypost.com/2021/09/06/federal-government-using-social-media-giants-to-censor-americans/

16 Olafimihan Oshin, "Twitter Bans Conservative Author Alex Berenson," *The Hill*, 29 August 2021. https://thehill.com/homenews/media/569908-twitter-bans-conservative-author-Alex-Berenson/

17 Emily S. Vogel, Monica Anderson, Margaret Porteus, et al., "Americans and

'Cancel Culture': Where Some See Calls for Accountability, Others See Censorship, Punishment," *Pew Research Center*, 19 May 2021. https://www.pewresearch.org/internet/2021/05/19/americans-and-cancel-culture-where-some-see-calls-for-accountability-others-see-censorship-punishment/

18 "Pennsylvania Assembly: Reply to the Governor, 11 November 1755," *Votes and Proceedings of the House of Representatives*, 1755–1756 (1756), pp. 19–21. https://founders.archives.gov/documents/Franklin/01-06-02-0107

19 Nick Andrews, Julia Stowe, Freja Kirsebom, et al., "Covid-19 Vaccine Effectiveness Against the Omicron (B.1.1.529) Variant," *New England Journal of Medicine*, 386: 1532–1546, 21 April 2022. https://www.nejm.org/doi/full/10.1056/NEJMoa2119451

CHAPTER 5

1 Jeffrey D. Sachs and Neil L. Harrison, "Questions Surrounding the Origins of COVID-19 Remain Unanswered," *Boston Globe*, 31 May 2022. https://www.bostonglobe.com/2022/05/31/opinion/questions-surrounding-origins-covid-19-remain-unanswered/

2 "Biological Weapons Convention," *United Nations Office for Disarmament Affairs*, accessed on 17 June 2022. https://www.un.org/disarmament/biological-weapons/

3 "Adherence to and Compliance with Arms Control, Nonproliferation, Disarmament Agreements, and Commitments," *U.S. Department of State*, August 2019, p. 45. https://www.state.gov/wp-content/uploads/2019/08/Compliance-Report-2019-August-19-Unclassified-Final.pdf

4 David Nakamura, Carol D. Leonnig, and Ellen Nakashima, "Matthew Pottinger Faced Communist China's Intimidation as a Reporter. Now He's at the White House Shaping Trump's Hard Line Policy Toward Beijing," *Washington Post*, 29 April 2020. https://www.washingtonpost.com/politics/matthew-pottinger-faced-communist-chinas-intimidation-as-a-reporter-hes-now-at-the-white-house-shaping-trumps-hard-line-policy-toward-beijing/2020/04/28/5fb3f6d4-856e-11ea-ae26-989cfce1c7c7_story.html

5 Josh Rogin, "State Department Cables Warned of Safety Issues In the Wuhan Lab Studying Bat Coronaviruses," *Washington Post*, 14 April 2020. https://www.washingtonpost.com/opinions/2020/04/14/state-department-cables-warned-safety-issues-wuhan-lab-studying-bat-coronaviruses/?fbclid=IwAR3B7IYoB7d1ZQK28Y74IXTshFBPEwl5sZxDmwlMqSF4wWZ--TMZaIxtnqE

CHAPTER 6

1 Bill Gertz, "Coronavirus Link to China Biowarfare Program Possible, Expert Says," *Washington Times*, 26 January 2020. https://www.washingtontimes.com/news/2020/jan/26/coronavirus-link-to-china-biowarfare-program-possi/?fbclid=IwAR2SobWLJgqGWoXeTDoGjSftzTKF6bBeenCKSzW6d7wK65butbxfvCKeA3M

NOTES

Natalie O'Neill, "WHO Scientist Wants Closer Look at Wuhan Lab That Moved Days Before COVID-19 Outbreak," *New York Post*, 13 August 2021. https://nypost.com/2021/08/13/who-scientist-eyes-on-wuhan-lab-that-moved-before-pandemic/amp/

3   Kevin Kunzmann, "WHO, China Report Suggests COVID-19 Passed from Bats to Humans Through Another Animal," *Contagion Live*, 29 March 2021. https://www.contagionlive.com/view/who-china-report-covid-19-passed-bats-humans-animal

4   Botao Xiao and Lei Xiao, "The Possible Origins of 2019-nCoV Coronavirus," *South China University of Technology*, 6 February 2020. https://s.rfi.fr/media/display/2f7b52c0-87b0-11ea-b8a0-005056bff430/Xiao%20Botao%20-%20The%20possible%20origins%20of%20the%202019-nCoV%20virus.pdf

5   David Ignatius, "How Did COVID-19 Begin? Its Initial Origin Story is Shaky," *Washington Post*, 2 April 2020. https://www.washingtonpost.com/opinions/global-opinions/how-did-covid-19-begin-its-initial-origin-story-is-shaky/2020/04/02/1475d488-7521-11ea-87da-77a8136c1a6d_story.html

6   Ross Ibbetson, "Did Coronavirus Originate in Chinese Government Laboratory? Scientists Believe Killer Disease May Have Begun in Research Facility 300 Yards from Wuhan Wet Fish Market," *Daily Mail*, 16 February 2020. https://www.dailymail.co.uk/news/article-8009669/Did-coronavirus-originate-Chinese-government-laboratory.html?fbclid=IwAR1nml3Tbjmie2jt_SCgUoprGLb_EN5YMyftq9cM-UfEwH0-3LdoZNzoum8

7   Aylin Woodward, "An Unsubstantiated Theory Suggests the Coronavirus Accidentally Leaked from a Chinese Lab – Here are the Facts," *Business Insider*, 15 April 2020. https://www.businessinsider.com/theory-coronavirus-accidentally-leaked-chinese-lab-2020-4

8   Tristin Hopper, "The (Very Strong) Case for COVID-19 Leaking from a Chinese Lab," *National Post*, 28 May 2021. https://nationalpost.com/news/the-very-strong-case-for-covid-19-leaking-from-a-chinese-lab

9   Helen Raleigh, "Coronavirus and China's Missing Citizen Journalists," *National Review*, 6 April 2020. https://www.nationalreview.com/magazine/2020/04/06/coronavirus-and-chinas-missing-citizen-journalists/

10 Dr. Embarek made his comments in a 2021 Danish television documentary entitled "The Virus Mystery."

11 Natalie O'Neill, "COVID-19 Lab Leak Theory a 'Probable Hypothesis,' WHO Scientist Says in Stunning Reversal," *New York Post*, 12 August 2021. https://nypost.com/2021/08/12/covid-19-lab-leak-theory-probable-hypothesis-who-scientist/

12 Adam Taylor, Emily Rauhala, and Martin Selsoe Sorensen, "In New Documentary, WHO Scientist Says Chinese Officials Pressured Investigation to Drop Lab-Leak Hypothesis," *Washington Post*, 12 August 2021. https://www.washingtonpost.com/world/2021/08/12/who-origins-embarek/

13 Ross McGuiness, "WHO Responds to Claims Wuhan Lab Worker Could Be

COVID Patient Zero," *Yahoo! News*, 13 August 2021. https://nz.news.yahoo.
com/who-wuhan-lab-worker-covid-patient-zero-120940428.html

14 Hinnerk Feldswich-Drentrup, "How the WHO Became China's Coronavirus Accomplice," *Foreign Policy*, 2 April 2020. https://foreignpolicy.com/2020/ 04/02/china-coronavirus-who-health-soft-power/

15 Alice Miranda Ollstein, "Trump Halts Funding to the World Health Organization," *Politico*, 14 April 2020. https://www.politico.com/news/2020/04/14/ trump-world-health-organization-funding-186786

16 Jamey Keaton, "Biden's US Revives Support for WHO, Reversing Trump Retreat," *AP News*, 21 January 2021. https://apnews.com/article/us-who-support-006ed181e016afa55d4cea30af236227

17 Lin Yang, "Pandemic Exposes Perils of Global Reliance on China for Drug Supplies," *VOA*, 19 May 2020. https://www.voanews.com/a/science-health_pandemic-exposes-perils-global-reliance-china-drug-supplies/6189571. html

18 Glen Owen, "WHO Chief 'Believes COVID Did Leak from Wuhan Lab' After a 'Catastrophic Accident' in 2019 Despite Publicly Maintaining 'All Hypotheses Remain on the Table,'" *Daily Mail*, 18 June 2022. https://www. dailymail.co.uk/news/article-10930501/WHO-chief-believes-Covid-DID-leak-Wuhan-lab-catastrophic-accident-2019.html?fbclid=IwAR0V2YOf WlMsZycNi5aHDacEkHgG1BpiTcwPen97uyUCES5IFBt3d52kuqU

19 John Ratcliffe, "John Ratcliffe: China Olympics 2022 – COVID Cover Up By Country's Leaders Means They Should Forfeit Games," *Fox News*, 2 August 2021. https://www.foxnews.com/opinion/china-olympics-2022-covid-cover-up-games-john-ratcliffe

20 Olivia Reingold, "Pompeo Insists COVID-19 Leaked from a Chinese Lab," *Politico*, 13 June 2021. https://www.politico.com/news/2021/06/13/pompeo-covid-chinese-lab-493986

21 Jane Perry, "Breaches of Safety Regulations are Probable Cause of Recent SARS Outbreak, WHO Says," *BMJ*, 328 (7450): 1222, 25 May 2004. https:// www.ncbi.nlm.nih.gov/pmc/articles/PMC416634/

22 David Cyranoski, "Inside the Chinese Lab Poised to Study World's Most Dangerous Pathogens," *Nature*, 23 February 2017. https://www.nature.com/ articles/nature.2017.21487

23 John Xie, "Chinese Lab with Checkered Safety Record Draws Scrutiny Over COVID-19," *Voice of America*, 21 April 2020. https://www.voanews.com/a/ covid-19-pandemic_chinese-lab-checkered-safety-record-draws-scrutiny-over-covid-19/6187947.html

24 Ibid.

25 David Cyranoski, "Bat Cave Solves Mystery of Deadly SARS Virus – and Suggests New Outbreak Could Occur," *Nature*, 1 December 2017. https:// www.nature.com/articles/d41586-017-07766-9

26 Ibid., "Chinese Lab with Checkered Safety Record Draws Scrutiny Over COVID-19."

27 Katsuji Nakazawa, "China Knew of Lab Safety Concerns from Last Year," *Nikkei Asia*, 30 April 2020. https://asia.nikkei.com/Editor-s-Picks/China-up-close/China-knew-of-lab-safety-concerns-from-last-year

28 Susana K.P. Lau, Patrick C.Y. Woo, Kenneth S.M. Li, and Kwok-Yung Yuen, "Severe Acute Respiratory Syndrome Coronavirus-Like Virus in Chinese Horseshoe Bats," *PNAS*, 102 (39), 14040–14045, 16 September 2005. https://www.pnas.org/doi/10.1073/pnas.0506735102

29 Ibid.

30 Jane Qiu, "How China's 'Bat Woman' Hunted Down Viruses from SARS to the New Coronavirus," *Scientific American*, 1 June 2020. https://www.scientificamerican.com/article/how-chinas-bat-woman-hunted-down-viruses-from-sars-to-the-new-coronavirus1/

31 Ibid.

32 James Walker, "Mike Pompeo Claims Intel Officials 'Didn't Want to Talk' About Wuhan Lab Leak Theory," *Newsweek*, 4 June 2021. https://www.newsweek.com/mike-pompeo-intel-officials-wuhan-lab-leak-cover-1597494

33 Alison Young, "Deleted COVID-19 Genetic Fingerprints Show It's Still Possible to Dig for Lab Leak Evidence," *USA Today*, 24 June 2021. https://www.usatoday.com/story/opinion/2021/06/24/covid-19-lab-leak-investigation-deleted-genetic-fingerprints-show-its-still-possible-dig-lab-leak-ev/7778194002/

34 Chan, Alina. Twitter post. June 22, 2021, 5:47 pm. https://twitter.com/ayjchan/status/1407455333115645956

35 Chan, Alina. Twitter post. March 1, 2021, 9:41 am. https://twitter.com/Ayjchan/status/1366398314791514118

36 Ibid.

37 Elsa B. Kania, "Minds at War: China's Pursuit of Military Advantage Through Cognitive Science and Biotechnology," *Prism*, 8, No. 3, accessed on 24 September 2022. https://ndupress.ndu.edu/Portals/68/Documents/prism/prism_8-3/prism_8-3_Kania_82-101.pdf

38 "Transcript: Matt Pottinger on 'Face the Nation,' February 21, 2021," *CBS News*, 21 February 2021. https://www.cbsnews.com/news/transcript-matt-pottinger-on-face-the-nation-february-21-2021/

39 Jocelyn Kaiser, "NIH Says Grantee Failed to Report Experiment in Wuhan That Created a Bat Virus That Made Mice Sicker," *Science*, 21 October 2021. https://www.science.org/content/article/nih-says-grantee-failed-report-experiment-wuhan-created-bat-virus-made-mice-sicker

40 Paul D. Thacker, "COVID-19: Lancet Investigation Into Origin of Pandemic Shuts Down Over Bias Risk," *BMJ*, 2021; 375:n2414. https://www.bmj.com/content/375/bmj.n2414

41 Alison Young and Jessica Blake, "Near Misses at UNC Chapel Hill's High Security Lab Illustrate Risk of Accidents with Coronaviruses," *ProPublica*, 17 August 2020. https://www.propublica.org/article/near-misses-at-unc-

chapel-hills-high-security-lab-illustrate-risk-of-accidents-with-coronaviruses

42 "UNC Project-China," *UNC School of Medicine*, accessed on 27 August 2022. https://globalhealth.unc.edu/china/

43 Mohana Basu, "Before Wuhan Row, How US-China Created SARS-Like Virus in 2015 to Show Its Pandemic Potential," *The Print*, 1 June 2021. https://theprint.in/science/before-wuhan-row-how-us-china-created-sars-like-virus-in-2015-to-show-its-pandemic-potential/668891/

44 Vineet D. Menachery, Boyd L. Yount, Jr., Kari Debbink, et al., "A SARS-Like Cluster Circulating Bat Coronaviruses Shows Potential for Human Emergence," *Nature*, 9 November 2015. https://www.nature.com/articles/nm.3985

45 Jon Cohen, "Prophet in Purgatory," *Science*, 17 November 2021. https://www.science.org/content/article/we-ve-done-nothing-wrong-ecohealth-leader-fights-charges-his-research-helped-spark-covid-19

CHAPTER 7

1 Trevor Scobey, Boyd L. Yount, Amy C. Sims, and Ralph S. Baric, "Reverse Genetics with a Full-Length Infectious cDNA of the Middle East Respiratory Syndrome Coronavirus," *PNAS*, 13 August 2013. https://www.pnas.org/doi/10.1073/pnas.1311542110

2 Rowan Jacobsen, "Inside the Risky Bat-Virus Engineering That Links America to Wuhan," *MIT Technology Review*, 29 June 2021. https://www.technologyreview.com/2021/06/29/1027290/gain-of-function-risky-bat-virus-engineering-links-america-to-wuhan/

3 Bryan Walsh, "The World Is Not Ready for the Next Pandemic," *Time*, 4 May 2017. https://time.com/magazine/us/4766607/may-15th-2017-vol-189-no-18-u-s/

4 David Gillum and Rebecca Moritz, "Why Gain-of-Function Research Matters," *The Conversation*, accessed on 20 June 2022. https://theconversation.com/why-gain-of-function-research-matters-162493

5 "Press Statement on the NSABB Review of H5N1 Research," *NIH News*, 20 December 2011. https://www.ncbi.nlm.nih.gov/books/NBK206979/

6 Jocelyn Kaiser, "EXCLUSIVE: Controversial Experiments That Could Make Bird Flu More Risky Poised to Resume," *Science*, 8 February 2019. https://www.science.org/content/article/exclusive-controversial-experiments-make-bird-flu-more-risky-poised-resume

7 Arturo Casadevall and Thomas Shenik, "The H5N1 Moratorium and Debate," *mBio*, No. 3, Vol. 5, 9 October 2012. https://journals.asm.org/doi/full/10.1128/mBio.00379-12

8 Wang Ning, Li Shi-Yue, Yang Xing-Lou, et al., "Serological Evidence of Bat SARS-Related Coronavirus Infection in Humans, China," *Virologica Sinica*, 21 November 2017. https://www.ecohealthalliance.org/wp-content/uploads/2018/03/Virologica-Sinica-SARSr.pdf

NOTES

9   Ibid.

10 Jef Akst, "Lab-Made Coronavirus Triggers Debate," *The Scientist*, 16 November 2015. https://www.the-scientist.com/news-opinion/lab-made-coronavirus-triggers-debate-34502

11 Savio Rodrigues, "COVID-19 Secret is with Baric, Daszak, and Zhengli," *Sunday Guardian Live*, 11 September 2021. https://www.sundayguardianlive.com/news/covid-19-secret-baric-daszak-zhengli

12 "Reconstruction of the 1918 Influenza Pandemic Virus," *CDC*, accessed on 21 June 2022. https://www.cdc.gov/flu/about/qa/1918flupandemic.htm

13 Noah Weber, "In Taiwan, Is It 'COVID-19' or 'Wuhan Pneumonia'?" *Sup China*, 6 April 2020. https://supchina.com/2020/04/06/in-taiwan-is-it-covid-19-or-wuhan-pneumonia/

14 David Cyranoski, "Profile of a Killer: The Complex Biology Powering the Coronavirus Pandemic," *Nature*, 4 May 2020. https://www.nature.com/articles/d41586-020-01315-7

15 Vaishali Basu Sharma, "Sars-cov2 is a Chimera with HIV Gene Manipulation," *PPF*, accessed on 21 June 2022. https://ppf.org.in/opinion/sars-cov2-is-a-chimera-with-hiv-gene-manipulation

16 Prashant Pradhan, Ashutosh Kumar Pandey, Akhilesh Mishra, et al., "Uncanny Similarity of Unique Inserts in the 2019-nCoV Spike Protein to HIV-1 gp120 and Gag," *bioRxiv*, accessed on 21 June 2022. https://www.biorxiv.org/content/10.1101/2020.01.30.927871v1

17 Katharine Lang, "Another Approved Malaria Medicine Shows Potential Against COVID-19," *Medical News Today*, 17 December 2021. https://www.medicalnewstoday.com/articles/another-approved-malaria-medicine-shows-potential-against-covid-19

18 Dave Makichuk, "French Prof Sparks Furor with Lab Leak Claim," *Asia Times*, 18 April 2020. https://asiatimes.com/2020/04/french-prof-sparks-furor-with-lab-leak-theory/

19 Luc Montagnier, Emilio Del Giudice, Jamal Aissa, et al., "Transduction of DNA Information Through Water and Electromagnetic Waves," *Pub Med*, 2015; 34(2): 106–12. https://pubmed.ncbi.nlm.nih.gov/26098521/

20 Heidi Medford, "Luc Montagnier (1932–2022)," *Nature*, 4 March 2022. https://www.nature.com/articles/d41586-022-00653-y

21 Beverly Rubik and Robert R. Brown, "Evidence for a Connection Between Coronavirus Disease-19 and Exposure to Radiofrequency Radiation from Wireless Communications Including 5G," *Pub Med*, 2021 October 26; 7(5): 666–681. https://www.ncbi.nlm.nih.gov/pmc/articles/PMC8580522/

22 Awaad K. Al Sarkhi, "The Link Between Electrical Properties of COVID-19 and Electromagnetic Radiation," *Biotechnology to Combat COVID-19*, 14 March 2021. https://www.intechopen.com/chapters/75714

23 "U of G Chemists Find Microwaves May Help Treat COVID-19," *University of Guelph News*, 28 February 2022. https://news.uoguelph.ca/2022/02/u-of-g-chemists-find-microwaves-may-help-treat-covid-19/

NOTES

1   "Coronavirus (COVID-19) Deaths," *Our World Data*, accessed on 25 September 2022. https://ourworldindata.org/covid-deaths

2   Stephanie Kum, Jennifer Kates, and Adam Wexler, "Economic Impact of COVID-19 on PEPFAR Countries," *Global Health Policy*, 7 February 2022. https://www.kff.org/global-health-policy/issue-brief/economic-impact-of-covid-19-on-pepfar-countries/

3   Bernd Debusmann, Jr., "COVID-19 Pandemic is Over in the US – Joe Biden," *BBC*, 20 September 2022. https://www.bbc.com/news/world-us-canada-62959089

4   Jacob Stern, "Fauci Addresses 'The Pandemic is Over,'" *The Atlantic*, 22 September 2022. https://www.theatlantic.com/health/archive/2022/09/fauci-addresses-the-pandemic-is-over/671507/

5   Jan Phillips, "Big Pharma Makes Big Bucks On Vaccines at Taxpayer Expense," *Durango Herald*, 7 August 2021. https://www.durangoherald.com/articles/big-pharma-makes-big-bucks-on-vaccines-at-taxpayer-expense/

6   Dominic Rushe, "The Richest Americans Became 40% Richer During the Pandemic," *The Guardian*, 5 October 2021. https://www.theguardian.com/media/2021/oct/05/richest-americans-became-richer-during-pandemic

7   Joanna Glasner, "Where Google, One of the Most Active Health Care Investors, Puts Its Capital," *Crunchbase*, 20 December 2021. https://news.crunchbase.com/startups/google-health-care-startups-investment-under-the-hood/

8   Todd Weatherby and Kelli Jonakin, "Executive Conversations: Accelerating COVID-19 Vaccine Development with Marcello Damiani, Chief Digital and Operational Excellence Officer at Moderna," *AWS*, 1 March 2021. https://aws.amazon.com/blogs/industries/executive-conversations-accelerating-covid-19-vaccine-development-with-marcello-damiani-chief-digital-and-operational-excellence-officer-at-moderna/

9   David Rhew, "Successful COVID-19 Vaccine Delivery Requires Strong Tech Partnerships," *Official Microsoft Blog*, 11 December 2020. https://blogs.microsoft.com/blog/2020/12/11/successful-covid-19-vaccine-delivery-requires-strong-tech-partnerships/

10 Jamie Nimmo, "Zuckerberg Fund Builds $100m Stake in Biotech Company Working with GlaxoSmithKline on Coronavirus Cures," *This is Money*, 30 May 2020. https://www.thisismoney.co.uk/money/news/article-8372251/Zuckerberg-fund-bets-100m-coronavirus-cures.html

11 CNN, "Thiel: Google Has $50B, Doesn't Innovate," *YouTube*, 17 July 2012. https://www.youtube.com/watch?v=2Q26XIKtwXQ

12 IIEA, "Eamonn Fingleton – The U.S. In the Coming Era of Chinese Dominance – 29 April 2014," *YouTube*, 14:08, April 30, 2014. https://www.youtube.com/watch?v=IdN_thfIiEg

NOTES

13 Declan Butler, "Engineered Bat Virus Stirs Debate Over Risky Research," *Nature*, 12 November 2015. https://www.nature.com/articles/nature.2015.18787

CHAPTER 9

1 "Scientists in China Discussed Weaponising Coronavirus: 2015 Report," *Business Standard*, 9 May 2021. https://www.business-standard.com/article/current-affairs/scientists-in-china-discussed-weaponising-coronavirus-in-2015-report-121050900916_1.html

2 "COVID: Biden Orders Investigation into Virus Origin as Lab Leak Theory Debated," *BBC*, 27 May 2021. https://www.bbc.com/news/world-us-canada-57260009

3 Ana Fifield, "'Wolf Warrior' Strives to Make China First with Coronavirus Vaccine," *Washington Post*, 5 March 2022. https://www.washingtonpost.com/world/asia_pacific/chinas-wolf-warrior-strives-to-be-first-with-coronavirus-vaccine/2020/03/19/d6705cba-699c-11ea-b199-3a9799c54512_story.html

4 New China TV, "'People's Hero' Chen Wei: Military Medical Scientist Working on Vaccine," *YouTube*, 12 September 2020. https://www.youtube.com/watch?v=1sqDWi0Jn7w

5 Sharad S. Chauhan, "Covid 19: The Chinese Military and Maj Gen Chen Wei," *Indian Defence Review*, 9 October 2020. http://www.indiandefencereview.com/spotlights/covid-19-the-chinese-military-and-maj-gen-chen-wei/

6 Angus Liu, "China's CanSino Bio Advances COVID-19 Vaccine into Phase 2 on Preliminary Safety Data," *Fierce Pharma*, 10 April 2020. https://www.fiercepharma.com/vaccines/china-s-cansino-bio-advances-covid-19-vaccine-into-phase-2-preliminary-safety-data

7 Anonymous OSINT Contributor, Billy Bostickson, and Giles Demaneuf, "An Investigation Into the WIV Databases That Were Taken Offline," *ResearchGate*, February 2021. https://www.researchgate.net/publication/349073738_An_investigation_into_the_WIV_databases_that_were_taken_offline

8 "Coronavirus: Why Have Two Reporters in Wuhan Disappeared?" *BBC*, 14 February 2020. https://www.bbc.com/news/world-asia-china-51486106

9 "Chinese Inhalable COVID-19 Booster Vaccine Safe and Effective: Study," *Xinhua*, 24 May 2022. https://english.news.cn/20220524/6b767673ac8047b69dba9d7d8f4dd355/c.html

10 Ian Bremmer, "Why the Chinese and Russian Vaccines Haven't Been the Geopolitical Wins They Were Hoping For," *Time*, 2 August 2021. https://time.com/6086028/chinese-russian-covid-19-vaccines-geopolitics/

11 Sofia Moutinho, "Chinese COVID-19 Vaccine Maintains Protection in Variant-Plagued Brazil," *Science*, 9 April 2021. https://www.science.org/content/article/chinese-covid-19-vaccine-maintains-protection-variant-plagued-brazil

12 Steven W. Mosher, "China's Deception Over COVID-19's Origins Gets More Outrageous Every Day," *New York Post*, 27 March 2021. https://nypost.com/2021/03/27/chinas-deception-over-covid-origins-more-outrageous-every-day/

13 Canrong Wu, Mengzhu Zheng, Yueying Yang, et al., "Furin: A Potential Therapeutic Target for COVID-19," *iScience*, Volume 23, Issue 10, October 2020. https://www.sciencedirect.com/science/article/pii/S2589004 220308348#!

14 "Trump Accuses China of 'Raping' US with Unfair Trade Policy," *BBC*, 2 May 2016. https://www.bbc.com/news/election-us-2016-36185012

15 Ibid.

16 Ibid.

17 "Trump-Taiwan Call Breaks US Policy Stance," *BBC*, 3 December 2016. https://www.bbc.com/news/world-us-canada-38191711

18 Ben Blanchard and Steve Holland, "China to Return Seized U.S. Drone, Says Washington 'Hyping Up' Incident," *Reuters*, 16 December 2016. https://www.reuters.com/article/us-usa-china-drone/china-to-return-seized-u-s-drone-says-washington-hyping-up-incident-idUSKBN14526J

19 George Friedman, *The Next 100 Years: A Forecast for the 21st Century* (New York: Random House, 2009), 50–64.

20 China Power Team, "How is China Feeding Its Population of 1.4 Billion?" *China Power*, 25 January 2017. https://chinapower.csis.org/china-food-security/

21 T. X. Hammes, "Offshore Control: A Proposed Strategy for an Unlikely Conflict," *Strategic Forum: National Defense University*, June 2012. https://ndupress.ndu.edu/Portals/68/Documents/stratforum/SF-278.pdf

22 Mackubin Owens, "Countering China's Grand Strategy," *American Greatness*, 28 November 2021. https://amgreatness.com/2021/11/28/countering-chinas-grand-strategy/

23 C.Y. Cyrus Chu and Ronald D. Lee, "Famine, Revolt, and the Dynastic Cycle," *Journal of Population Economics*, 7, 351–378, 1994. https://link.springer.com/article/10.1007/BF00161472

24 Jason Gambill, "China and Russia are Waging Irregular Warfare Against the United States: It Is Time for a U.S. Global Response, Led by Special Operations Command," *Joint Intermediate Force Capabilities Office*, 15 November 2021. https://jnlwp.defense.gov/Press-Room/In-The-News/Article/2857039/china-and-russia-are-waging-irregular-warfare-against-the-united-states-it-is-t/

25 Robert D. Blackwill and Jennifer M. Harris, *War by Other Means: Geoeconomics and Statecraft* (Cambridge: Harvard University Press, 2016), pp. 93–128.

26 Brandon J. Weichert, "Much More Than a Trade War with China," *New English Review*, June 2019. https://www.newenglishreview.org/articles/much-more-than-a-trade-war-with-china/

27 David P. Goldman, "Trade Wars Part Two – The Empire Strikes Back," *Asia*

Times, 6 August 2019. https://asiatimes.com/2019/08/trade-wars-part-two-the-empire-strikes-back/

28 Wendong Zhang, Lulu Rodriguez, and Shuyang Qu, "3 Reasons Midwest Farmers Hurt by the U.S.-China Trade War Still Support Trump," *The Conversation*,4November2019.https://theconversation.com/3-reasons-midwest-farmers-hurt-by-the-u-s-china-trade-war-still-support-trump-126303

29 Bloomberg Economics. Twitter post. 22 December 2019. 8:31pm. https://twitter.com/economics/status/1208923079734439936?ref_src=twsrc%5 Etfw%7Ctwcamp%5Etweetembed%7Ctwterm%5E1208923079734439936 %7Ctwgr%5E%7Ctwcon%5Es1_&ref_url=https%3A%2F%2Fwww.dailywire. com%2Fnews%2Famid-trump-trade-war-china-lowers-import-tariffs-on-over-850-products%3Futm_term%3Dutm_campaign%3Ddw_conversions_subscriptions_performancemax_politicalutm_source%-3Dadwordsutm_medium%3Dppchsa_acc%3D6411461344hsa_cam%3D165 99826472hsa_grp%3Dhsa_ad%3Dhsa_src%3Dxhsa_tgt%3Dhsa_kw%3 Dhsa_mt%3Dhsa_net%3Dadwordshsa_ver%3D3gclid%3DCjoKCQjwnt CVBhDdARIsAMEwAClZRw5mxCdluy7kQ95mjUhZ3qoc1UeUiF1bB2O j2hB9gSKT92TiTWoaAiKzEALw_wcB

30 Elliot Smith, "China's Surging Food Prices Won't Weaken Its Hand in the Trade War, Economists Say," *CNBC*, 9 August 2019. https://www.cnbc.com/ 2019/08/09/chinas-surging-food-prices-wont-weaken-its-hand-in-the-trade-war.html

31 Elsa B. Kania and Lorand Laskai, "Myths and Realities of China's Civil-Military Fusion Strategy," *CNAS*, 28 January 2021. https://www.cnas.org/ publications/reports/myths-and-realities-of-chinas-military-civil-fusion-strategy

32 Death By China, "Death By China: How America Lost Its Manufacturing Base (Official Version)," *YouTube*, 10 April 2016, 14:29–14:42. https://www. youtube.com/watch?v=mMlmjXtnIXI

33 Ibid., Chauhan, "Covid 19: The Chinese Military and Maj Gen Chen Wei," http://www.indiandefencereview.com/spotlights/covid-19-the-chinese-military-and-maj-gen-chen-wei/

34 Dany Shoham, "About the Genomic Origin and Direct Source of the Pandemic Virus," *Journal of Defense Studies*, Vol. 16, No. 2, April-June 2022, pp. 79–92. https://www.idsa.in/system/files/jds/jds-16-2-2022-dany-shoham_compressed.pdf

35 Ibid.

36 Ibid.

37 David Gilbert, "A Chinese Doctor Injected Herself with an Untested Coronavirus Vaccine," *Vice*, 4 March 2020. https://www.vice.com/en/article/ v74p5y/a-chinese-doctor-injected-herself-with-an-untested-coronavirus-vaccine

38 Jim Wiesemeyer, "Who Actually Won the U.S., China Trade War?" *AgWeb*

Farm Journal, 23 May 2022. https://www.agweb.com/news/policy/politics/ who-actually-won-us-china-trade-war

39 Michael Collins, David Jackson, and John Fritze, "U.S., China Sign 'Phase One' Trade Agreement, Signaling Pause in Trade War," *USA Today*, 16 January 2020. https://www.usatoday.com/story/news/politics/2020/01/15/ trump-sign-china-trade-deal-details-remain-mystery/4434592002/

40 David P. Goldman, "China Suppressed Covid-19 with AI and Big Data," *Asia Times*, 3 March 2020. https://asiatimes.com/2020/03/china-suppressed-covid-19-with-ai-and-big-data/

41 Rebecca Fannin, "The Rush to Deploy Robots in China Amid the Coronavirus Outbreak," *CNBC*, 2 March 2020. https://www.cnbc.com/2020/03/02/ the-rush-to-deploy-robots-in-china-amid-the-coronavirus-outbreak.html

42 Lily Kuo, "China Becomes First Major Economy to Recover from Covid-19 Pandemic," *The Guardian*, 19 October 2020. https://www.theguardian.com/ business/2020/oct/19/china-becomes-first-major-economy-to-recover-from-covid-19-pandemic

43 Jonathan Cheng, "China is the Only Major Economy to Report Economic Growth for 2020," *Wall Street Journal*, 18 January 2021. https://www.wsj. com/articles/china-is-the-only-major-economy-to-report-economic-growth-for-2020-11610936187

44 Bruno Maçães, "How China Could Become the World's Largest Economy Much Sooner Than Expected," *Big Think*, 7 March 2022. https://bigthink. com/the-future/geopolitics-for-the-end-time/

45 Brandon J. Weichert, "How Covid Made the Present World," *Asia Times*, 23 June 2021. https://asiatimes.com/2021/06/how-covid-made-the-present-world/

46 Jianli Yang and Aaron Rhodes, "How China Has Crushed Hong Kong's Democracy," *National Review*, 18 March 2021. https://www.nationalreview. com/2021/03/how-china-has-crushed-hong-kongs-democracy/

47 "Grassley, Johnson Release Bank Records Tying Biden Family to CCP-Linked Individuals and Companies," *Chuck Grassley*, 29 March 2022. https://www.grassley.senate.gov/news/remarks/grassley-johnson-release-bank-records-tying-biden-family-to-ccp-linked-individuals-and-companies

48 Xiao Zibang, "Russia Overtakes Saudi Arabia as China's Top Oil Producer," *Bloomberg*, 20 June 2022. https://www.bloomberg.com/news/articles/2022-06-20/china-buys-7-5-billion-of-russian-energy-with-oil-at-record

49 Aine Quinn and Megan Durisin, "Big Win for Russian Wheat on Go Ahead to Sell More to China," *Bloomberg*, 4 February 2022. https://www.bloomberg.com/news/articles/2022-02-04/big-win-for-russian-wheat-as-china-further-opens-its-market#xj4y7vzkg

50 Adam Quinn, "The Art of Declining Politely: Obama's Prudent Presidency and the Waning of American Power," *International Affairs*, Vol. 87, No. 4, July 2011. https://www.jstor.org/stable/20869760

51 Carol Leonnig and Philip Rucker, *I Alone Can Fix It: Donald J. Trump's Catastrophic Final Year* (Penguin Press: New York, 2021), p. 110.

52 Jack Feehan, "Is COVID-19 the Worst Pandemic?" *Mauritas*, 149: 56–58, July 2021. https://www.ncbi.nlm.nih.gov/pmc/articles/PMC7866842/

53 Gregory Copley, *The New Total War of the 21st Century and the Trigger of the Fear Pandemic* (Alexandria: ISSA, 2020), pp. 247–49.

54 Tyrone Clarke, "Scott Morrison Demands Answers Over COVID Origins After WHO Wuhan Lab Leak Concession," *Sky News*, 16 July 2021. https://www.skynews.com.au/australia-news/politics/scott-morrison-demands-answers-over-covid-origins-after-who-wuhan-lab-leak-concession/news-story/b0f36abb176d41fb2c5d77d3846cbf79

55 Natasha Kassam, "Great Expectations: The Unraveling of the Australia-China Relationship," *Brookings Institute*, 20 July 2020. https://www.brookings.edu/articles/great-expectations-the-unraveling-of-the-australia-china-relationship/

56 Mark Kortepeter, "A Defense Expert Explores Whether the COVID-19 Coronavirus Makes a Good Bioweapon," *Forbes*, 21 August 2020. https://www.forbes.com/sites/coronavirusfrontlines/2020/08/21/a-defense-expert-explores-whether-the-covid-19-coronavirus-makes-a-good-bioweapon/?sh=132545837ece

CHAPTER 10

1 Justin Ling, "A Brilliant Scientist Was Mysteriously Fired from a Winnipeg Virus Lab. No One Knows Why," *MacLean's*, 15 February 2022. https://www.macleans.ca/longforms/winnipeg-virus-lab-scientist/

2 Ibid.

3 Robert Fife and Steven Chase, "RCMP Investigating Winnipeg Scientists Fired from Lab for Possible Transfer of Intellectual Property to China," *The Globe and Mail*, 30 June 2021. https://www.theglobeandmail.com/politics/article-scientists-fired-from-winnipeg-lab-under-rcmp-investigation-for/

4 Annie Melchor, "Canadian Official Reprimanded for Withholding Winnipeg Lab Info," *The Scientist*, 23 June 2021. https://www.the-scientist.com/news-opinion/canadian-official-reprimanded-for-withholding-winnipeg-lab-info-68919

5 Karen Pauls, "'Wake-Up Call for Canada': Security Experts Say Case of 2 Fired Scientists Could Point to Espionage," *CBC*, 10 June 2021. https://www.cbc.ca/news/canada/manitoba/winnipeg-lab-security-experts-1.6059097

6 "Bioterrorism Agents/Diseases," *Centers for Disease Control*, accessed on 24 June 2022. https://emergency.cdc.gov/agent/agentlist-category.asp

7 Ibid., Ling, https://www.macleans.ca/longforms/winnipeg-virus-lab-scientist/

8 Robert Fife, "Chinese Major-General Worked with Fired Scientist at Canada's Top Infectious Disease Lab," *The Globe and Mail*, 16 September 2021. https://

www.theglobeandmail.com/politics/article-chinese-pla-general-collaborated-with-fired-scientist-at-canadas-top/

9   Karen Pauls and Kimberly Ivany, "Mystery Around 2 Fired Scientists Points to Larger Issues at Canada's High-Security Lab, Former Colleagues Say," *CBC*, 8 July 2021. https://www.cbc.ca/news/canada/manitoba/nml-scientists-speak-out-1.6090188

10   Tenzin Zompa, "Chinese Military's Epidemiologist Worked with Fired Scientist at Canada's Top Disease Lab: Report," *The Print*, 17 September 2021. https://theprint.in/world/chinese-militarys-epidemiologist-worked-with-fired-scientist-at-canadas-top-disease-lab-report/734940/

11   Ibid., Ling.

12   Ibid.

13   K. Lloyd Billingsley, "Covering for Dr. Xiangguo Qiu: 'Nothing to Steal' from Canada's National Lab?" *Independent Institute*, 14 July 2021. https://www.independent.org/news/article.asp?id=13660

14   Robert Fife, Steven Chase, and Shannon Vanraes, "Whereabouts of Fired Winnipeg Scientists at Centre of National-Security Investigation Still Unclear," *Globe and Mail*, 22 June 2022. https://www.theglobeandmail.com/politics/article-fired-winnipeg-scientists-location-unclear/

15   Committee on Science, Space, and Technology, "Scholars or Spies: Foreign Plots Targeting America's Research and Development," *115th Congress*, 11 April 2018. https://www.govinfo.gov/content/pkg/CHRG-115hhrg29781/pdf/CHRG-115hhrg29781.pdf

16   Alsamman M. Alsamman and Hatem Zayed, "The Transcriptomic Profiling of SARS-CoV-2 Compared to SARS, MERS, EBOV, and H1N1," *PLOS One*, 10 December 2020. https://journals.plos.org/plosone/article?id=10.1371/journal.pone.0243270

17   Ibid. "'Wake-Up Call for Canada': Security Experts Say Case of 2 Fired Scientists Could Point to Espionage," https://www.cbc.ca/news/canada/manitoba/winnipeg-lab-security-experts-1.6059097

CHAPTER 11

1   "About Tom," *Tom Cotton Senate for Arkansas*, accessed on 22 June 2022. https://www.cotton.senate.gov/about

2   Alex Isenstadt, "Cotton Gathers Big Donors to Talk 2024 Presidential Race," *Politico*, 14 June 2022. https://www.politico.com/news/2022/06/14/cotton-big-donors-2024-presidential-race-00039476

3   Tom Cotton, "Coronavirus and the Laboratories in Wuhan," *Wall Street Journal*, 21 April 2020. https://www.wsj.com/articles/coronavirus-and-the-laboratories-in-wuhan-11587486996?mod=opinion_lead_pos5

4   "About the Committee," *U.S. Senate Select Committee on Intelligence*, accessed on 22 June 2022. https://www.intelligence.senate.gov/about

5   Chris Cillizza, "The Story Keeps Getting Worse for Andrew Cuomo on

NOTES

COVID-19," *CNN*, 12 February 2021. https://www.cnn.com/2021/02/12/politics/andrew-cuomo-nursing-homes-covid-19/index.html

6   "COVID-19 Deaths in Wuhan Seem Far Higher Than the Official Count," *The Economist*, 30 May 2021. https://www.economist.com/graphic-detail/2021/05/30/covid-19-deaths-in-wuhan-seem-far-higher-than-the-official-count

7   Yi Rao, "My Relatives in Wuhan Survived. My Uncle in New York Did Not," *New York Times*, 22 July 2020. https://www.nytimes.com/2020/07/22/opinion/coronavirus-china-us.html

8   Zachary Evans, "NYT Publishes Op-Ed by Chinese Professor Without Disclosing Ties to Beijing," *National Review*, 23 July 2020. https://www.nationalreview.com/news/nyt-publishes-op-ed-by-chinese-professor-who-mocked-trump-by-calling-aids-american-sexually-transmitted-disease-in-post-on-propaganda-outlet/

9   Charles Creitz, "China 'Uses Elite Capture to Pay Off' US Oligarchs Instead of 'Head-to-Head' Conflict,'" *Fox News*, 16 February 2022. https://www.foxnews.com/media/china-elite-american-business-corporations-biden-schweizer

10   Tom Cotton, "Coronavirus Lab-Leak-Theory Proponents Have Been Vindicated," *National Review*, 8 June 2021. https://www.nationalreview.com/2021/06/coronavirus-lab-leak-theory-proponents-have-been-vindicated/

11   Steven W. Mosher, "Here's All the Proof Biden Needs to Conclude COVID-19 Was Leaked from a Lab," *New York Post*, 24 July 2021. https://nypost.com/2021/07/24/heres-all-the-proof-biden-needs-to-conclude-covid-19-was-leaked-from-a-lab/

12   "Statement by President Joe Biden on the Investigation into the Origins of COVID-19," *The White House*, 26 May 2021. https://www.whitehouse.gov/briefing-room/statements-releases/2021/05/26/statement-by-president-joe-biden-on-the-investigation-into-the-origins-of-covid-19/

CHAPTER 12

1   Warren Fiske, "Fact-Check: Did Fauci Say Coronavirus was 'Nothing to Worry About'?" *PolitiFact*, 29 April 2020. https://www.statesman.com/story/news/politics/elections/2020/04/29/fact-check-did-fauci-say-coronavirus-was-nothing-to-worry-about/984113007/

2   "Proclamation on Suspension of Entry as Immigrants and Nonimmigrants of Persons Who Pose a Risk of Transmitting 2019 Novel Coronavirus," *The White House*, 31 January 2020. https://trumpwhitehouse.archives.gov/presidential-actions/proclamation-suspension-entry-immigrants-nonimmigrants-persons-pose-risk-transmitting-2019-novel-coronavirus/

3   Newsmax, "Dr. Anthony Fauci's Thoughts on COVID-19 from January 2020." *Twitter*, 6 April 2020. 11:12 AM. https://twitter.com/newsmax/status/1247180304823062529?lang=en

4   The Cats Roundtable, "Dr. Anthony Fauci 1-26-20," *SoundCloud*, 25 January 2020. https://soundcloud.com/john-catsimatidis/dr-anthony-fauci-1-26-20

5   Conversations with Dr. Bauchner, "Coronavirus Infections – More Than Just the Common Cold," *JN Learning*, 23 January 2020. https://edhub. ama-assn.org/jn-learning/audio-player/18197306

6   "President Trump with Coronavirus Task Force Briefing," *C-SPAN*, 9 March 2020. 33:53. https://www.c-span.org/video/?470172-1/president-trump-coronavirus-task-force-briefing

7   Gregg Re, "Coronavirus Timeline Shows Politicians', Media's Changing Rhetoric on Risk of Pandemic," *Fox News*, 20 April 2020. https://www.fox-news.com/politics/from-new-york-to-canada-to-the-white-house-initial-coronavirus-responses-havent-aged-well

8   KPIX CBS SF Bay Area, "Speaker Pelosi Visits SF's Chinatown to Show Support Amid Coronavirus Fears," *YouTube*, 24 February 2020. https://www.youtube.com/watch?v=eFCzoXhNM6c

9   NYC Mayor's Office. *Twitter.* 13 February 2020. 6:16 pm. https://twitter.com/NYCMayorsOffice/status/1228095506368344066

10 The Aspen Institute, "Ron Klain on Coronavirus," *YouTube*, 20 March 2020. https://www.youtube.com/watch?v=h8KZ3F7JwTo

11 PTI, "Trump's Move to Ban Travel Between US and China Not a Great Moment: House Speaker Nancy Pelosi," *The New Indian Express*, 27 April 2020. https://www.newindianexpress.com/world/2020/apr/27/trumps-move-to-ban-travel-between-us-and-china-not-a-great-moment-house-speaker-nancy-pelosi-2135693.html

12 NBC Bay Area Staff, "Nancy Pelosi Visits San Francisco's Chinatown Amid Coronavirus Concerns," *NBC*, 24 February 2020. https://www.nbcbayarea.com/news/local/nancy-pelosi-visits-san-franciscos-chinatown/2240247/

13 Alexandra Kelley, "Fauci: Why the Public Wasn't Told to Wear Masks When the Coronavirus Pandemic Began," *The Hill*, 16 June 2020. https://thehill.com/changing-america/well-being/prevention-cures/502890-fauci-why-the-public-wasnt-told-to-wear-masks/

14 Nathaniel Weixel, "Federal Stockpile of Emergency Medical Equipment Depleted, House Panel Says," *The Hill*, 8 April 2020. https://thehill.com/policy/healthcare/491871-federal-stockpile-of-emergency-medical-equipment-depleted-house-panel-says/

15 Ken Roberts, "China More Dominant Than Ever in Covid-Related 'PPE' – And U.S. Flags," *Forbes*, 19 September 2020. https://www.forbes.com/sites/kenroberts/2020/09/19/china-more-dominant-than-ever-in-covid-related-ppe---and-us-flags/?sh=35444d6117f7

16 Paul Bond, "How Americans' Opinion of Dr. Anthony Fauci Has Changed Over the Past Year," *Newsweek*, 2 June 2021. https://www.newsweek.com/how-americans-opinion-dr-anthony-fauci-has-changed-over-past-year-1596690

NOTES

17 Bruce Schreiner, "Sen. Rand Paul Wants to Investigate Origins of COVID-19," *AP News*, 30 April 2022. https://apnews.com/article/2022-midterm-elections-covid-health-kentucky-anthony-fauci-1e6e7828c2133754ce641f794b10379d

18 Lincoln Mitchell, "Dr. Fauci and Sen. Rand Paul's Recent Tense Exchange Over Covid Sends a Message," *NBC*, 12 January 2022. https://www.nbcnews.com/think/opinion/dr-fauci-sen-rand-paul-s-recent-tense-exchange-over-ncna1287366

19 Chris Riotta, "From HIV to Covid-19: Fauci on His 'Complicated Relationship' with Activist Larry Kramer," *NBC*, 2 October 2020. https://www.nbcnews.com/feature/nbc-out/hiv-covid-19-dr-fauci-his-complicated-relationship-larry-kramer-n1241684

20 Libby Cathey and Sasha Pezenki, "Fauci, Rand Paul Get Into a Shouting Match Over Wuhan Lab Research," *ABC*, 20 July 2021. https://abcnews.go.com/Politics/fauci-rand-paul-shouting-match-wuhan-lab-research/story?id=78946568

21 Robert Towey, "Fauci Says Rand Paul is 'Egregiously Incorrect' About Gain of Function Research in Senate Showdown," *CNBC*, 4 November 2021. https://www.cnbc.com/2021/11/04/fauci-says-rand-paul-egregiously-incorrect-about-gain-of-function-research.html

22 Ed Browne, "Fauci Was 'Untruthful' to Congress About Wuhan Lab Research, New Documents Appear to Show," *Newsweek*, 9 September 2021. https://www.newsweek.com/fauci-untruthful-congress-wuhan-lab-research-documents-show-gain-function-1627351

23 Nora Colburn, "How Can We Know the COVID-19 Vaccine Won't Have Long-Term Side Effects?" *The Ohio State University Wexner Medical Center*, 14 September 2021. https://wexnermedical.osu.edu/blog/covid-19-vaccine-long-term-side-effects

24 Jon Cohen, "'Absolutely Remarkable': No One Who Got Moderna's Vaccine in Trial Developed Severe COVID-19," *NBC*, 30 November 2020. https://www.science.org/content/article/absolutely-remarkable-no-one-who-got-modernas-vaccine-trial-developed-severe-covid-19

25 MacKenzie Sigalos, "You Can't Sue Pfizer or Moderna If You Have Severe Covid Vaccine Side Effects. The Government Likely Won't Compensate You for Damages Either," *CNBC*, 23 December 2020. https://www.cnbc.com/2020/12/16/covid-vaccine-side-effects-compensation-lawsuit.html

26 Adam Andrzejewski, "Rand Paul is Doing the Right Thing by Asking for Transparency in the NIH: Opinion," *Courier-Journal*, 20 June 2022. https://www.courier-journal.com/story/opinion/2022/06/20/rand-pauls-questioning-dr-faucis-nih-royalty-payments-right/7622469001/

27 Ibid.

28 Ibid.

29 C-Span, "Heated Exchange Between Sen. Rand Paul & Dr. Anthony Fauci on Vaccines and Royalties," *YouTube*, 16 June 2022. https://www.youtube.com/watch?v=3ICBBK-d-Co

30 Ibid.

31 Ibid.

32 Ibid.

33 Ibid.

34 According to the *New England Journal of Medicine*, "The Bayh-Dole Act of 1980 permitted scientists, universities, and businesses to patent and profit from discoveries made through federally funded research." Howard Markel, "Patents, Profits, and the American People – The Bayh-Dole Act of 1980," *New England Journal of Medicine*, 29 August 2013. https://www.nejm.org/doi/full/10.1056/nejmp1306553

35 Ibid., C-Span.

36 Adam Andrzejewski, "Substack Investigation: Fauci's Royalties and the $350 Million Royalty Payment Stream HIDDEN by NIH," *Open the Books*, 16 May 2022. https://www.openthebooks.com/substack-investigation-faucis-royalties-and-the-350-million-royalty-payment-stream-hidden-by-nih/

37 D'Angelo Gore, "Some Posts About NIH Royalties Omit Fauci Statement That He Donates His Payments," *Fact Check*, 20 May 2022. https://www.factcheck.org/2022/05/scicheck-some-posts-about-nih-royalties-omit-that-fauci-said-he-donates-his-payments/

38 Tae-Wook Chun, Danielle Murray, Jesse S. Justement, et al., "Broadly Neutralizing Antibodies Suppress HIV in the Persistent Viral Reservoir," *Proc Natl Acad Sci USA*, 9 September 2014, 111(36): 13151–13156. https://www.ncbi.nlm.nih.gov/pmc/articles/PMC4246957/

39 "The Institute of Human Virology: 2019 Annual Report," *The University of Maryland School of Medicine*, p. 14, accessed on 26 June 2022. https://www.ihv.org/media/SOM/Microsites/IHV/documents/Annual-Report/2019-IHV-Annual-Report-final-v7-low-res.pdf

40 Ibid., p. 30.

41 Klara K. Eriksson, Divine Makia, Reinhard Maier, et al., "Towards a Coronavirus-Based HIV Multigene Vaccine," *Clin Dev Immunol.*, 13(2–4): 353–60, Jun-Dec 2006. https://pubmed.ncbi.nlm.nih.gov/17162377/

42 "Scientists Are Working on HIV Vaccines Based on COVID Vaccine Tech," *Science Friday*, 17:08, 1 April 2022. https://www.sciencefriday.com/segments/fauci-hiv-vaccine-covid-mrna/

43 Zhujun Ao, Lijun Wang, J. Mendoza, Keding Cheng, et al., "Incorporation of Ebola Glycoprotein into HIV Targeting Particles Facilitates Dendritic Cell and Macrophage Targeting and Enhances HIV-Specific Immune Responses," *PLOS ONE*, 17 May 2019. https://journals.plos.org/plosone/article?id=10.1371/journal.pone.0216949

CHAPTER 13

1 Albert Einstein, "Why Socialism?" *Monthly Review*, May 1949. https://monthlyreview.org/2009/05/01/why-socialism/

NOTES

2   "Oppenheimer Security Hearing," *Atomic Heritage Foundation*, 7 July 2014. https://www.atomicheritage.org/history/oppenheimer-security-hearing

3   "Klaus Fuchs," *Atomic Heritage Foundation*, accessed on 27 June 2022. https://www.atomicheritage.org/profile/klaus-fuchs

4   Robert D. McFadden, "David Greenglass, the Brother Who Doomed Ethel Rosenberg, Dead at 92," *New York Times*, 14 October 2014. https://www.nytimes.com/2014/10/15/us/david-greenglass-spy-who-helped-seal-the-rosenbergs-doom-dies-at-92.html

5   Oleg Yegorov, "Why Did the Rosenbergs Spy and Die for Communism?" *Russia Beyond*, 16 March 2018. https://www.rbth.com/history/327844-why-did-the-rosenbergs-spy-for-the-ussr

6   Claudia Deane, Kim Parker, and John Gramlich, "A Year of U.S. Public Opinion on the Coronavirus Pandemic," *Pew Research Center*, 5 March 2021. https://www.pewresearch.org/2021/03/05/a-year-of-u-s-public-opinion-on-the-coronavirus-pandemic/

7   Jack Brewster, "Nearly Half of Americans Believe COVID-19 Leaked from a Lab, Poll Finds," *Forbes*, 9 June 2021. https://www.forbes.com/sites/jackbrewster/2021/06/09/nearly-half-of-americans-believe-covid-19-leaked-from-chinese-lab-poll-finds/?sh=152e65906431

8   Adam Frank, "What is Scientism, and Why is It a Mistake?" *Big Think*, 9 December 2021. https://bigthink.com/13-8/science-vs-scientism/

9   Yonat Shimron, "Poll: Americans' Belief in God is Dropping," *Washington Post*, 24 June 2022. https://www.washingtonpost.com/religion/2022/06/24/poll-americans-belief-god-is-dropping/

10 "Scientists, Politics, and Religion," *Pew Research Center*, 9 July 2009. https://www.pewresearch.org/politics/2009/07/09/section-4-scientists-politics-and-religion/

11 Giuliana Viglione, "Scientists Strongly Back Joe Biden for US President in Nature Poll," *Nature*, 23 October 2020. https://www.nature.com/articles/d41586-020-02963-5

12 Andrew Isaac Meso, "Another Diversity Problem – Scientists' Politics," *Nature*, 8 December 2020. https://www.nature.com/articles/d41586-020-03479-8?WT.ec_id=NATURE-20201210&utm_source=nature_etoc&utm_medium=email&utm_campaign=20201210&sap-outbound-id=A3144545E3EB42B49FA04221AE0D04B3DA89D153

13 Robert A. Sirico, "The Role of Responsibility in a Free Society," *Acton Institute*, 20 July 2010. https://www.acton.org/pub/religion-liberty/volume-8-number-4/role-responsibility-free-society

CHAPTER 14

1   Qiang Zha, "Can China Reverse the Brain Drain?" *University World News*, 28 March 2014. https://www.universityworldnews.com/post.php?story=20140326132305490

2   Hepeng Jia, "What is China's Thousand Talents Plan?" *Nature Jobs Career*

Guide, 2018. https://media.nature.com/original/magazine-assets/d41586-018-00538-z/d41586-018-00538-z.pdf

3   Bill Gertz, "China's Intelligence Networks in United States Include 25,000 Spies," *Washington Free Beacon*, 11 July 2017. https://freebeacon.com/national-security/chinas-spy-network-united-states-includes-25000-intelligence-officers/

4   "The Spy Next Door?" *Ash Clinical News*, June 2019. https://ashpublications.org/ashclinicalnews/news/4599/The-Spy-Next-Door

5   Ibid., "What is China's Thousand Talents Plan?" https://media.nature.com/original/magazine-assets/d41586-018-00538-z/d41586-018-00538-z.pdf

6   Christopher Burgess, "China's Thousand Talents Program Harvests U.S. Technology and a Guilty Verdict," *Clearance Jobs*, 22 December 2021. https://news.clearancejobs.com/2021/12/22/chinas-thousand-talents-program-harvests-u-s-technology/

7   "The China Threat: Chinese Talent Plans Encourage Trade Secret Theft, Economic Espionage," *Federal Bureau of Investigation*, accessed on 29 June 2022. https://www.fbi.gov/investigate/counterintelligence/the-china-threat/chinese-talent-plans

8   Nathaniel Benchley, "The $24 Swindle," *American Heritage*, December 1959. https://www.americanheritage.com/24-swindle

9   Robert Pear, "U.S. Officials Warn Health Researchers: China May Be Trying to Steal Your Data," *New York Times*, 6 January 2019. https://www.nytimes.com/2019/01/06/us/politics/nih-china-biomedical-research.html

10 Francis M. Collins, "Dear Colleagues Letter," *Department of Health and Human Services*, 20 August 2018. https://www.insidehighered.com/sites/default/server_files/media/NIH%20Foreign%20Influence%20Letter%20to%20Grantees%2008-20-18.pdf

CHAPTER 15

1   Jack Beyrer, "Ohio State Researcher Sentenced to Prison for Concealing Ties to Chinese Spy Program," *Washington Free Beacon*, 14 May 2021. https://freebeacon.com/national-security/ohio-state-researcher-sentenced-to-prison-for-concealing-ties-to-chinese-spy-program/

2   Center for Development of Security Excellence, "Case Study: Fraud/Theft of Intellectual Property: Song Guo Zheng," *Defense Counterintelligence and Security Agency*, accessed on 30 June 2022. https://www.cdse.edu/Portals/124/Documents/casestudies/case-study-zheng.pdf

3   "University Researcher Sentenced to Prison for Lying on Grant Applications to Develop Scientific Expertise for China," *Department of Justice*, 14 May 2021. https://www.justice.gov/opa/pr/university-researcher-sentenced-prison-lying-grant-applications-develop-scientific-expertise

4   Ibid.

NOTES

CHAPTER 16

1   Mara Hvistendahl, "Exclusive: Major U.S. Cancer Center Ousts 'Asian' Researchers After NIH Flags Their Foreign Ties," *Science*, 19 April 2019. https://www.science.org/content/article/exclusive-major-us-cancer-center-ousts-asian-researchers-after-nih-flags-their-foreign

2   "The Chinese Consulate in Houston May Have Provided Financial and Logistical Support to Protest Groups," *The Liberty Web*, 8 September 2020. https://eng.the-liberty.com/2020/8025/

3   "Chinese Consulate in Houston Ordered to Close by US," *BBC*, 23 July 2020. https://www.bbc.com/news/world-us-canada-53497193

4   Amanda Macias, "FBI Arrests Chinese Researcher for Visa Fraud After She Hid at Consulate in San Francisco," *CNBC*, 24 July 2020. https://www.cnbc.com/2020/07/24/chinese-researcher-arrested-after-she-hid-at-consulate-in-san-francisco.html

5   Associated Press, "U.S. Moves to Drop Visa Fraud Charge Against a Chinese Researcher," *NPR*, 23 July 2021. https://www.npr.org/2021/07/23/1019680651/us-china-visa-fraud-case-charges

6   "National Cancer Moonshot Initiative," *American Association for Cancer Research*, accessed on 28 August 2022. https://www.aacr.org/professionals/policy-and-advocacy/science-policy-government-affairs/national-cancer-moonshot-initiative/

7   Damian Garde, "Biden Maps Out a 'Moonshot' Approach to Cancer with Plans to 'Break Down Silos' in R&D," *Fierce Biotech*, 13 January 2016. https://www.fiercebiotech.com/r-d/biden-maps-out-a-moonshot-approach-to-cancer-plans-to-break-down-silos-r-d

8   Ben Adams, "Chinese Scientists Become First to Test CRISPR in Humans, as 'Sputnik 2.0' Begins," *Fierce Biotech*, 16 November 2016. https://www.fiercebiotech.com/biotech/chinese-scientists-become-first-to-test-crispr-humans-as-sputnik-2-0-begins

9   You Lu, Jianxin Xue, Tony Mok, et al., "Safety and Feasibility of CRISPR-Edited Cells in Patients with Refractory Non-Small-Cell Lung Cancer," *Nature Medicine*, 26, 732–40, 27 April 2020. https://www.nature.com/articles/s41591-020-0840-5

10 Rasmus Kragh Jakobsen, "First Chinese CRISPR Gene Therapy Trial Demonstrates Safety," *Crispr News Medicine*, 28 April 2020. https://crisprmedicinenews.com/news/first-chinese-crispr-gene-therapy-trial-demonstrates-safety/

11 Zhi Zong, Yujun Wei, Jiang Ren, et al., "The Intersection of COVID-19 and Cancer: Signaling Pathways and Treatment Implications," *Molecular Cancer*, 20, 76, 2021. https://molecular-cancer.biomedcentral.com/articles/10.1186/s12943-021-01363-1

12 Lei Mao, Yu Zhang, and Leaf Huang, "mRNA Vaccine for Cancer

Immunotherapy," *Molecular Cancer*, 41, 25 February 2021. https://molecu-lar-cancer.biomedcentral.com/articles/10.1186/s12943-021-01335-5

CHAPTER 17

1   "NIH, DARPA, and FDA Collaborate to Develop Cutting-Edge Technolo-gies to Predict Drug Safety," *National Institutes of Health*, 16 September 2011. https://www.nih.gov/news-events/news-releases/nih-darpa-fda-collaborate-develop-cutting-edge-technologies-predict-drug-safety

2   Research and Markets, "Global Market for Bioelectronic Medicine to 2025 – Key Role That Government Funding Agencies Such as the NIH and DARPA are Playing in the Development of Industry," *CISION*, 14 June 2019. https://www.prnewswire.com/news-releases/global-market-for-bio electronic-medicine-to-2025---key-role-that-government-funding-agencies-such-as-the-nih-and-darpa-are-playing-in-the-development-of-the-industry-300867735.html

3   Robert Cook-Deegan, "Does NIH Need a DARPA?" *Issues in Science and Technology*, Vol. 13, No. 2, Winter 1997. https://issues.org/national-institutes-health-nih-darpa-cook-deegan/

4   "Advanced Research Projects Agency for Health (ARPA-H): Congressional Action and Selected Policy Issues," *Congressional Research Service*, 15 April 2022, p. 2. https://sgp.fas.org/crs/misc/R47074.pdf

5   Terry Magnuson, "NIH On a Path to Add a DARPA-Like Model," *UNC Research*, 1 July 2021. https://research.unc.edu/2021/07/01/nih-on-a-path-to-add-a-darpa-like-model/

6   Nathan Blouin, Email Message to Ralph S. Baric, et al., May 17, 2017. https://usrtk.org/wp-content/uploads/2021/12/UNC_Baric_12.30.21.pdf

7   Natasha Gilbert and Max Kozlov, "The Controversial China Initiative is Ending – Researchers are Relieved," *Nature*, 24 February 2022. https://www.nature.com/articles/d41586-022-00555-z

8   Nidhi Subbaraman, "Scientists' Fears of Racial Bias Surge Amid US Crack-down on China Ties," *Nature*, 29 October 2021. https://www.nature.com/articles/d41586-021-02976-8

9   Edward Wong, "A Chinese Empire Reborn," *New York Times*, 5 January 2018. https://www.nytimes.com/2018/01/05/sunday-review/china-military-eco-nomic-power.html

CHAPTER 18

1   John Mack, "Nanotechnology: What's in It for Biotech?" *Biotechnology Healthcare*, 2(6): 29–61, 35–36, December 2005. https://www.ncbi.nlm.nih.gov/pmc/articles/PMC3571017/

2   "What is Nanotechnology?" *National Nanotechnology Institute*, accessed on 30 June 2022. https://www.nano.gov/nanotech-101/what/definition

3   Ibid.

NOTES

4   Sonia Contera, Jorge Bernardino de la Serna, and Teresa D. Tetley, "Biotechnology, Nanotechnology, and Medicine," *Emerging Topics in Life Science*, 4, 6, pp. 551–54, 9 December 2020. https://portlandpress.com/emergtoplifesci/article/4/6/551/227204/Biotechnology-nanotechnology-and-medicine

5   Rumina Tenchov, "Understanding the Nanotechnology in COVID-19 Vaccines," *CAS*, 18 February 2021. https://www.cas.org/resource/blog/understanding-nanotechnology-covid-19-vaccines

6   "Charles M. Lieber Named University Professor," *The Harvard Gazette*, 20 July 2017. https://news.harvard.edu/gazette/story/2017/07/chemist-charles-m-lieber-receives-harvards-highest-faculty-honor/

7   Ellen Barry, "In a Boston Court, a Superstar of Science Falls to Earth," *New York Times*, 21 December 2022. https://www.nytimes.com/2021/12/21/science/charles-lieber.html

8   "Harvard University Professor and Two Chinese Nationals Charged in Three Separate China Related Cases," *Department of Justice*, 28 January 2020. https://www.justice.gov/usao-ma/pr/harvard-university-professor-convicted-making-false-statements-and-tax-offenses

9   Jeffrey Mervis, "U.S. Prosecutor Leading China Probe Explains Effort to That Led to Charges Against Harvard Chemist," *Science*, 3 February 2020. https://www.science.org/content/article/us-prosecutor-leading-china-probe-explains-effort-led-charges-against-harvard-chemist

10 Ibid., *Department of Justice.*

11 Hindustan Times, "U.S. Convicts Harvard Prof for Lying About China Ties. Crackdown to Curb Beijing Influence in U.S.?" *YouTube*, 22 December 2021. https://www.youtube.com/watch?v=Igo8_wOrm6c

12 Brandon J. Weichert, "China's Quest for Exotic Tech: Brain Control Interface," *The Weichert Report*, 7 February 2021. https://theweichertreport.wordpress.com/2021/02/07/chinas-quest-for-exotic-tech-brain-control-interface/

13 NTD News Today, "China Pursues 'Brain Control' Weaponry," *NTD*, 11 January 2022. https://www.ntd.com/china-pursues-brain-control-weaponry_726017.html

14 Mahbube K. Siddiki, "China as the World Leader in Nanotechnology: Another Wakeup Call for the West," *Small Wars Journal*, 12 March 2022. https://smallwarsjournal.com/jrnl/art/china-world-leader-nanotechnology-another-wakeup-call-west

15 Ibid.

16 Jonathan Shaw, "Virus-Sized Transistors," *Harvard Magazine*, January-February 2011. https://www.harvardmagazine.com/2011/01/virus-sized-transistors

17 Ben Halder, "How China is the Future of Nanoscience," *Ozy*, 3 February 2020. https://www.ozy.com/the-new-and-the-next/cloning-to-cancer-china-is-driving-the-future-of-small-science/256094/

18 Ibid.

19 Ibid.

20 Ibid.

21 "Tianxia (All Under Heaven)," *Chinese Thought*, accessed on 1 July 2022. https://www.chinesethought.cn/EN/shuyu_show.aspx?shuyu_id=2161

CHAPTER 19

1   Richard Dobbs and Jaana Remes, "Introducing The Most Dynamic City of 2025," *Foreign Policy*, 13 August 2012. https://foreignpolicy.com/2012/08/13/introducing-the-most-dynamic-cities-of-2025/

2   "Spotlight on Wuhan," *Nature*, 13 May 2015. https://www.ncbi.nlm.nih.gov/pmc/articles/PMC7095288/

3   Ibid.

4   Ibid.

5   "The Future Trend of the World's New Military Revolution," *Reference News*, 24 August 2017. https://web.archive.org/web/20190823210313/http://www.xinhuanet.com/politics/2017-08/24/c_129687890.htm

6   Dany Shoham, "China's Biological Warfare Programme: An Integrative Study with Special Reference to Biological Weapons Capabilities," *Journal of Defence Studies*, Vol. 9, No. 2, April 2015. https://idsa.in/jds/9_2_2015_ChinasBiologicalWarfareProgramme

7   *The Asia Times*, "Webinar: You Will Be Assimilated – China's Plan to Sino-Form the World," *YouTube*, 30 September 2020. https://www.youtube.com/watch?v=IsZTqNCS7jA

CHAPTER 20

1   T. Randolph Beard, David L. Kaserman, and Rigmar Osterkamp, *The Global Organ Shortage: Economic Causes, Human Consequences, Policy Responses* (Stanford: Stanford University Press, 2013), pp. 92–112.

2   "Do Stem Cells Hold the Key for the Future of Transplantation?" *National Kidney Foundation*, accessed on 3 July 2022. https://www.kidney.org/news/newsroom/fs_new/stemcellskey

3   Erik Seedhouse, "The Human Clone Market," *Beyond Human*, 2 August 2014, pp. 51–64. https://www.ncbi.nlm.nih.gov/pmc/articles/PMC7122979/

4   Chi-Hun Park, Young-Hee Jeoung, Kyung-Jung uh, et al., "Extraembyonic Endoderm (XEN) Cells Capable of Contributing to Embryonic Chimeras Established from Pig Embryos," *Stem Cell Reports*, Vol. 16, Issue 1, 17 December 2020, pp. 212–23. https://www.cell.com/stem-cell-reports/fulltext/S2213-6711(20)30459-8?_returnURL=https%3A%2F%2Flinkinghub.elsevier.com%2Fretrieve%2Fpii%2FS2213671120304598%3Fshowall%3Dtrue

5   Ibid., Shoham, "China's Biological Warfare Programme."

6   David Cyranoski, "Gene-Edited 'Micropigs' to Be Sold as Pets at Chinese Institute," *Nature*, 526, 18, 29 September 2015. https://www.nature.com/articles/nature.2015.18448

7   "The Newest Members of Beijing's Police Force are 6 Cloned K9s," *CBS*

News, 21 November 2019. https://www.cbsnews.com/news/cloned-police-dogs-china-beijing-security-forces/

8 "Cloned Police Dogs are Getting a Lot of Attention at the Expo," *Sinogene*, accessed on 3 July 2022. https://www.sinogene.org/cloned-police-dogs-are-getting-a-lot-of-attention-at-the-expo.html

9 Antonio Regalado, "Pet Cloning is Bringing Human Cloning a Little Bit Closer," *MIT Technology Review*, 13 April 2018. https://www.technology review.com/2018/04/13/143901/human-cloning-just-got-a-little-bit-closer-heres-why/

10 Jasper Becker, "China's Mutant Monkeys: These are Just Two of the Countless Animals Used in Secret Genetic Engineering Tests in Labs – Many with Appalling Biosecurity. No Wonder So Many Experts Say Covid DID Leak from Wuhan Research Centre, Writes Jasper Becker," *Daily Mail*, 5 June 2021. https://www.dailymail.co.uk/news/article-9655357/JASPER-BECKER-No-wonder-experts-say-Covid-DID-leak-Wuhan-research-centre.html

11 Jack Dunhill, "This Artificial Womb and AI Nanny is the Future of Child Development, Chinese Scientists Claim," *IFL Science*, 31 January 2022. https://www.iflscience.com/this-artificial-womb-and-ai-nanny-is-the-future-of-child-development-claim-chinese-scientists-62437

12 John Ratcliffe, "China is National Security Threat No. 1," *Wall Street Journal*, 3 December 2020. https://www.wsj.com/articles/china-is-national-security-threat-no-1-11607019599

13 Reuters in Washington, "China the 'Greatest Threat to Democracy and Freedom,' U.S. Spy Chief Warns," *The Guardian*, 3 December 2020. https://www.theguardian.com/us-news/2020/dec/03/china-beijing-america-democracy-freedom

14 "NIH Research Involving Introduction of Human Pluripotent Cells into Non-Human Vertebrate Animal Pre-Gastrulation Embryos," *National Institutes of Health*, 23 September 2015. https://grants.nih.gov/grants/guide/notice-files/NOT-OD-15-158.html

15 Antonio Regalado, "Chinese Scientists Have Put Human Brain Genes in Monkeys – And, Yes, They May Be Smarter," *MIT Technology Review*, 10 April 2019. https://www.technologyreview.com/2019/04/10/136131/chinese-scientists-have-put-human-brain-genes-in-monkeysand-yes-they-may-be-smarter/

16 Lori Marino, "We've Created Human-Pig Chimeras – But We Haven't Weighed the Ethics," *Stat*, 26 January 2017. https://www.statnews.com/2017/01/26/chimera-humans-animals-ethics/

17 Sharon Begley, "First Human-Pig Chimeras Created, Sparking Hopes for Transplantable Organs – And Debate," *Stat*, 26 January 2017. https://www.statnews.com/2017/01/26/first-chimera-human-pig/

CHAPTER 21

1   Joshua Lipes, "Expert Says 1.8 Million Uyghurs, Muslim Minorities Held in Xinjiang's Internment Camps," *Radio Free Asia*, 24 November 2019. https://www.rfa.org/english/news/uyghur/detainees-11232019223242.html

2   Matthew Hill, David Campanale, and Joel Gartner, "'Their Goal is to Destroy Everyone': Uighur Camp Detainees Allege Systematic Rape," *BBC*, 2 February 2021. https://www.bbc.com/news/world-asia-china-55794071

3   Yael Grauer, "Revealed: Massive Chinese Police Database," *The Intercept*, 29 January 2021. https://theintercept.com/2021/01/29/china-uyghur-muslim-surveillance-police/

4   Charlotte Jee, "China is Using DNA Samples to Try to Re-Create the Faces of Uighurs," *MIT Technology Review*, 3 December 2019. https://www.technologyreview.com/2019/12/03/102429/china-is-using-dna-samples-to-try-to-recreate-the-faces-of-uighurs/

5   Ibid.

6   Beatrice Brown, "DNA Phenotyping Experiment on Uighurs Raises Ethical Questions About Informed Consent," *Bill of Health*, 9 December 2019. https://blog.petrieflom.law.harvard.edu/2019/12/09/dna-phenotyping-experiment-on-uighurs-raises-ethical-questions-about-informed-consent/

7   Volker Roelcke, "Nazi Medicine and Research on Human Beings," *The Lancet*, December 2004. https://www.thelancet.com/journals/lancet/article/PIIS0140-6736(04)17619-8/fulltext

8   Ben Westcott, "Chinese Media Calls for 'People's War' as US Trade War Heats Up," *CNN*, 14 May 2019. https://www.cnn.com/2019/05/14/asia/china-us-beijing-propaganda-intl

9   Leon Hadar, "Biden's Trump-Lite China Trade Policy," *Global Zeitgeist*, 12 October 2021. https://leonhadar.substack.com/p/bidens-trump-lite-china-trade-policy

CHAPTER 22

1   Kirsty Needham and Clare Baldwin, "China's Gene Giant Harvests Data from Millions of Women," *Reuters*, 7 July 2021. https://www.reuters.com/investigates/special-report/health-china-bgi-dna/

2   Ibid.

3   Ibid.

4   Ibid.

5   Andrea Park, "BGI Genomics Accused of Partnering with Chinese Military to Harvest DNA from Prenatal Tests: Reuters," *Fierce Biotech*, 9 July 2021. https://www.fiercebiotech.com/medtech/bgi-genomics-accused-partnering-chinese-military-to-harvest-data-from-prenatal-tests

6   Jocelyn Kaiser, "Genome Researchers Question Security Provisions In New U.S. Senate Bill," *Science*, 14 June 2021. https://www.science.org/content/article/genome-researchers-question-security-provisions-new-us-senate-bill

NOTES

7   Tom Cotton and Mike Gallagher, "Letter to Secretary of Treasury Janet Yellen, Secretary of Commerce Gina M. Raimondo, and Secretary of Defense Lloyd Austin," *Congress of the United States*, 28 September 2021. https://www.cotton.senate.gov/imo/media/doc/bgi_letter.pdf

8   Eric Schmidt, Bob Work, et al., "Final Report: National Security Commission on Artificial Intelligence," *National Security Commission on Artificial Intelligence*, February 2021. https://www.nscai.gov/wp-content/uploads/2021/03/Full-Report-Digital-1.pdf

9   Klon Kitchen and Bill Drexel, "Pull US AI Research Out of China," *Defense One*, 10 August 2021. https://www.defenseone.com/ideas/2021/08/pull-us-ai-research-out-china/184359/

10 Julian E. Barnes, "U.S. Warns of Efforts by China to Collect Genetic Data," *New York Times*, 22 October 2021. https://www.nytimes.com/2021/10/22/us/politics/china-genetic-data-collection.html

11 "China's Collection of Genomic and Other Healthcare Data from America: Risks to Privacy and U.S. Economic and National Security," *The National Counterintelligence and Security Center*, February 2021. https://www.dni.gov/files/NCSC/documents/SafeguardingOurFuture/NCSC_China_Genomics_Fact_Sheet_2021revision20210203.pdf

12 Steve Friess, "Concerns Over Chinese Genomics Bid," *Politico*, 4 December 2012. https://www.politico.com/story/2012/12/concerns-arise-in-chinese-bid-for-genomics-firm-084516

13 Andrew Hessel, Marc Goodman, and Steven Kotler, "Hacking the President's DNA," *The Atlantic*, November 2012. https://www.theatlantic.com/magazine/archive/2012/11/hacking-the-presidents-dna/309147/

CHAPTER 23

1   "Biohacking is Becoming the New DIY Activity," *BBVA*, 22 August 2017. https://www.bbva.com/en/biohacking-becoming-new-diy-activity/

2   "Biohacking Conference," *Biohacking Congress*, accessed on 31 August 2022. https://biohackingcongress.com

3   "Biohacking," *Dictionary*, accessed on 4 July 2022. https://www.merriam-webster.com/dictionary/biohacking

4   Brandon J. Weichert, "Understanding Digital Security," *Real Clear Public Affairs*, 19 August 2019. https://www.realclearpublicaffairs.com/articles/2019/08/19/understanding_digital_security_18786.html

5   Associated Press in Washington, "US Government Hack Stole Fingerprints of 5.6 Million Federal Employees," *The Guardian*, 23 September 2015. https://www.theguardian.com/technology/2015/sep/23/us-government-hack-stole-fingerprints

6   Dan Wang, "China Hawks Don't Understand How Science Advances," *The Atlantic*, 12 December 2021. https://www.theatlantic.com/ideas/archive/2021/12/china-initiative-intellectual-property-theft/621058/

7   Rory Truex, "What the Fear of China is Doing to American Science," *The

    Atlantic, 16 February 2021. https://www.theatlantic.com/ideas/archive/
    2021/02/fears-about-china-are-disrupting-american-science/618031/

8   Scott Moore, "China's Biotech Boom Could Transform Lives – Or Destroy
    Them," *Foreign Policy*, 8 November 2019. https://foreignpolicy.com/
    2019/11/08/cloning-crispr-he-jiankui-china-biotech-boom-could-
    transform-lives-destroy-them/

INDEX

DESIGN & COMPOSITION BY CARL W. SCABROUGH

First American edition published in 2023 by Encounter Books, an activity of Encounter for Culture and Education, Inc., a nonprofit, tax-exempt corporation. Encounter Books website address: www.encounterbooks.com

Manufactured in the United States and printed on acid-free paper. The paper used in this publication meets the minimum requirements of ANSI/NISO Z39.48—1992 (R 1997) (*Permanence of Paper*).

FIRST AMERICAN EDITION

LIBRARY OF CONGRESS CATALOGING-IN-PUBLICATION

Names: Weichert, Brandon J., 1988- author.
Title: Biohacked : China's race to control life / Brandon J. Weichert.
Description: First American edition. | New York, NY : Encounter Books, 2023. | Includes bibliographical references and index. |
Identifiers: LCCN 2022038061 (print) | LCCN 2022038062 (ebook) | ISBN 9781641773225 (hardcover) | ISBN 9781641773232 (ebook)
Subjects: LCSH: Biotechnology industries—Political aspects—China. | Biotechnology industries—Moral and ethical aspects—China. | Biotechnology—Research—Government policy—China. | China—Politics and government—2002–
Classification: LCC HD9999.B443 C6745 2023 (print) | LCC HD9999.B443 (ebook) | DDC 338.4/766060951—dc23/eng/20221103
LC record available at https://lccn.loc.gov/2022038061
LC ebook record available at https://lccn.loc.gov/2022038062

1 2 3 4 5 6 7 8 9   20 23